WHITE SHARK

A BIOGRAPHY OF THE FISH
THAT SCARED THE WORLD

WHITE
SHARK

MICHAEL BRIGHT

\Bᵇ\
Biteback Publishing

First published in Great Britain in 2025 by
Biteback Publishing Ltd, London
Copyright © Michael Bright 2025

ISBN 978-1-78590-930-6

10 9 8 7 6 5 4 3 2 1

A CIP catalogue record for this book is available from the British Library.

Set in Minion Pro and Futura PT

Printed and bound in Great Britain by
CPI Group (UK) Ltd, Croydon CR0 4YY

FSC
www.fsc.org
MIX
Paper | Supporting
responsible forestry
FSC® C013604

CONTENTS

INTRODUCTION

Imagine a fish that can be the length of a large automobile and weigh almost as much, that possesses jaws which have the strongest bite force of any living animal and are lined with row upon row of teeth that are, in effect, serrated knives and pin-sharp forks, and you will start to form a picture in your mind of the world's largest macropredatory shark – the infamous great white shark. Precise dimensions for the biggest individuals are hard to come by. There's always something wrong with the measuring technique and estimates are often wide of the mark. Even so, reliable figures from accurate measurements indicate the largest living individuals could be close to 6.5 metres (21 feet) long, that's four times as long as the average person is tall, and it's quite possible there are even bigger ones.

Scientists are able to work out the size of a shark even when it's swum away. They measure the size of the bite marks it's left behind, and they've come up with some astonishing results. Off southern Australia, for instance, bite marks on whale carcasses indicate white sharks in the region of 8 metres (26 feet) long, and off New Zealand a fisherman compared the length of a white shark he encountered with that of his boat and came up with an estimated figure of close to 9 metres (30 feet) – a real giant!

These monster white sharks are generally mature females, males being considerably shorter, and it's quite possible that they have lived to a ripe old age, more than seventy years old by some estimations, so they are not only the biggest predatory sharks, but possibly also among the oldest and wisest, adapted to an extraordinary degree to the watery environment in which they live.

To many people, this species is the ultimate hunter-killer. At various stages in its life, it hunts and kills quite different prey, adding more species to its diet as it gets older, including, sadly, people. It has been responsible for more shark bite incidents and human deaths than any other species, according to the International Shark Attack File of the University of Florida, probably because it sees us as aberrant marine mammals flailing about in the water and certainly worth an exploratory bite; but a gentle mouthing from a 2-tonne white shark could result in a bad mauling and severe damage to any victim who attracted its attention. Even so, if you consider how many of us pursue recreational pastimes in the sea – wading, swimming, surfing, paddleboarding, kayaking – the number of incidents is surprisingly few. If the shark actually saw us as food, there would be far more attacks and many more deaths, but the mythology associated with white sharks and humans remains, which means that people want to believe that it's a monster and man-eater.

The problem is that, down the years, the shark has gained an unsavoury reputation. Even though this notoriety was well known to ancient mariners, to whom the shark was known quite bluntly as 'man-eater' and 'white death', it was not until the mid-1970s that the wider public became aware of this now famous fish. On 20 June 1975, at 464 cinema screens across North America, the phenomenon that was *Jaws* was unleashed on an unsuspecting public. On 25 July, the total number increased to 675, in what was the first 'wide release' of a film. Usually, films are 'slow released', drip fed into a few cinemas in order to build up interest, but *Jaws*, directed by Steven Spielberg, went for broke, and it paid off, breaking box office records. It was the original 'summer blockbuster', so called because queues of people went around the block outside cinemas; previously films had tended not to be released at this time of year, when people were on holiday.

The launch built on the 'monster' nature of its principal non-human co-star. Billboards on the front of cinemas read 'The Terrifying No. 1 Bestseller – Now a Terrifying Motion Picture'. The bestseller was Peter Benchley's book of the same name published the year before, with 5.5 million copies sold in the USA alone by the time the film was released. White sharks didn't stand a chance.

The white shark's reputation after the *Jaws* phenomenon, of course, went to rock bottom, and the species suffered as a consequence. In California, for example, the film led to vendetta killings and great white shark tournaments (see also Chapter 12), and together with a commercial fishery, these almost completely wiped out the population of white sharks along the west coast of North America. A set of mounted white shark jaws became the ultimate shark angler's trophy, selling for up to $50,000 apiece, and hundreds of white sharks died as a consequence; but during the past fifty years, since the film's first release, the public interest in the shark, and even the fear of it, has sparked a renaissance in great white shark research.

The film may have terrorised people, but it also galvanised a new generation of shark researchers. In recent years, the great white is the shark species that has probably received the most scientific attention of any, leading almost to a softening of its character. Scientists even give tagged individuals affectionate names, all the result of the remarkable things that are being revealed about its life.

In fact, down the centuries, the white shark has had many names. 'Great white shark' or 'great white', probably on account of its size and a pure white belly, are the most popular monikers among an English-speaking general public today, while just plain 'white shark' seems to be the preferred English common name used by shark scientists. In Australia, it is known as the 'white pointer' (see below). Elsewhere in the world people have their own names in their own languages: in Spain it is *gran tiburón blanco* or *jaquetón blanco* meaning 'white jacket'; and in the Afrikaans language of South Africa it is called *witdoodhai* meaning 'white death shark'. The shark also has its own unique scientific name, and it too has a history.

The discipline of taxonomy – the science of naming, describing and classifying living things – made a great leap forward when the Swedish naturalist Carl Linnaeus introduced the binomial system of naming a species, for example *Homo sapiens* for modern humans. It was designed to enable scientists from all over the world to be able to talk about the same plant, animal or microbe and understand each other, rather than in the haphazard way it was done before using mainly common names that only locals understood.

In the tenth edition of his *Systema Naturae*, published in 1758, he described and classified the white shark, but it was not without its howlers. For starters, he grouped the shark with 'amphibians' and assigned it the scientific name *Squalus carcharias*, *Squalus* being the genus in which he placed all sharks, and *carcharias* used generally by the Greeks to describe either a 'point' or a 'type of shark'. The word 'point' in translation led to the white shark's common name in Australia being 'white pointer'. As it happened, 'carcharias' became established as a favoured specific epithet, but the universally accepted binomial species name came much later.

Indeed, the white shark gained many scientific names down the years as scientists failed to agree – *Squalus caninus* (1765), *Carcharias lamia* (1810) and *Carcharias vulgaris* (1836) to name just a few; in fact, it wasn't until 1838 that Sir Andrew Smith coined the genus name *Carcharodon*, from the Greek *karcharodōn*, meaning 'sharp and odious teeth', a reference to the white shark's rows of serrated, triangular teeth, and so eventually the white shark's full species name became *Carcharodon carcharias*, the scientific name it possesses to this very day.

Although it's the sole representative of its genus, the white shark is one of the mackerel sharks in the family Lamnidae, from the Greek word *lamna*, meaning 'fish of prey'. Its closest living relatives are three others in the family – the longfin and shortfin mako sharks, the latter being the fastest shark in the ocean, along with the salmon shark and the porbeagle, two sharks that closely resemble each other and are superficially white shark lookalikes. These two live in the Pacific and Atlantic oceans respectively.

All these mackerel shark characters have a fusiform shape, rounded and tapering at both ends, and are powerfully built, hydrodynamically very efficient and very fast, the ocean's apex predators. The caudal (tail) fin has the upper and lower lobes almost the same size, giving these sharks a crescent-shaped tail for high-speed swimming. One diagnostic feature is the keel or keels on the caudal peduncle, the narrow bit that joins the tail to the body. There is a single keel on either side of the caudal peduncle of the white shark, a feature that it shares with mako sharks, and which distinguishes them from other mackerel sharks, which have a main keel and two shorter secondary keels. It's an easy way to distinguish white sharks from salmon sharks and porbeagles. They all have a tall and stiff first dorsal fin and exceptionally large teeth, both iconic images that we tend to associate with the white shark.

White sharks are normally white below but dark on top, a phenomenon known as countershading, but just occasionally an oddball might appear. Leucism, in which all or part of an animal is abnormally white, is rare in white sharks, but on at least two occasions white sharks that are white all over have been reported. The first, a baby no more than 1.5 metres (5 feet) long, was caught off Boknesstrand in Eastern Cape, South Africa, in 1996; in fact, it was an albino, with characteristic pink eyes and a slightly yellow skin. The second was another young shark washed up on a sandy beach at Port Hacking, New South Wales, Australia. Unlike the South African shark, this specimen was white all over, with normal dark eyes, so it would be described as leucistic – in other words, an actual fully 'white shark'.

A normal white shark's eyes are often described as 'black', but that's because we usually see only the pupil. The iris is a deep blue, and, in keeping with all the mackerel sharks, the eyes of the white shark are much larger than those of other shark species of a comparable size, and it can see relatively well below and above the waves (see also Chapter 7).

All the sharks in the mackerel family generally prefer colder waters to warm, with occasional forays into the tropics. The white shark is found in preferred sea surface temperatures between 12 and 24°C (54 and 75°F), although tagged sharks have been known to enter waters as cold as 2.7°C (37°F) and as warm as 27°C (81°F).

So, where in the world do white sharks hang out? As it happens, they tend not to stay in one place, although they might return to a particular location time and again. They are accomplished ocean migrants, something that has only become apparent in recent years. They were always thought of as a coastal species but now we know that this is far from the truth. As adults, they spend the greater part of their life in the open ocean, returning inshore mainly to feed on high-energy foods, such as seal blubber, which is when we generally see them, hence the initial mistake (see also Chapter 6). Large numbers are found off the east coast of the USA and Canada, and the Cape Cod sharks are currently thought to be one of the largest concentrations of white sharks in the world, much to the chagrin of the region's beachgoers and, let's face it, the local populations of seals. They also inhabit the Gulf of Mexico and even waters around Caribbean islands, such as Cuba, where one of the world's largest white sharks was caught (see also Chapter 10). More recently, the Bahamas, especially the western edge of the Tongue of the Ocean, a deep-water basin close to Central Andros Island, has been known to hide them. The sharks pass through the area at a depth of 25 metres (82 feet) between dusk and dawn, when few people are about, so until a recent tagging programme revealed their nocturnal migration, nobody knew they were there at all.

North America's west coast is another area with large numbers of white sharks, the population having bounced back from the post-1975 onslaught in Californian waters. They occur from Mexico in the south to Alaska in the north, with one individual 6 metres (20 feet) long caught in the central Bering Sea in the Arctic, so the sharks here experience a broad range of sea surface temperatures and sea conditions. Individuals from this population of white sharks also frequent the Hawaii archipelago, in the middle of the Pacific Ocean, where long ago various species, such as tiger and white sharks, and Galápagos and grey reef sharks, were captured for gladiatorial contests. Long before it was a US Navy base, Pearl Harbor had a 1.6-hectare (4-acre) marine enclosure into which sharks were lured with bloody meat. When the gates were closed and the enclosure was filled with sharks, the contests began, each gladiatorial slave equipped with nothing more than a stick with a shark's tooth at the end with which to fight off their adversaries. Some won their freedom, others didn't.

In the southern hemisphere, key white shark centres include Western Australia, New South Wales and Queensland, where the population is stable, the count aided by the study of shark DNA (see also Chapter 12). Farther to the south are the New Zealand hotspots, including its sub-Antarctic islands, like Campbell Island, just 500 kilometres (320 miles) from the Antarctic coast and site of the southernmost white shark attack ever recorded.

In the Indian Ocean, the south-east coast of South Africa

is a known white shark centre, especially Gansbaai, meaning 'bay of geese', which has been home to a thriving cage diving industry. South Africa, Mozambique, Zanzibar and Kenya are on the same white shark migration route. White sharks are also present off Madagascar, along with the oceanic islands of Mauritius, the Seychelles and Réunion, the latter the site of several shark bite incidents. Bull and tiger sharks are among the perpetrators. In August 2006 a 34-year-old male had a leg severed by a white shark, but he drowned before he could be brought ashore. And, in 1975 and 1972 swimmers were killed by white sharks at beaches near the island's capital, Saint-Denis, and a tourist beach at Saint-Philippe; in fact, of fifty-six recorded shark attacks here, seventeen have been by white sharks. Elsewhere in the Indian Ocean, such as off Pakistan, white sharks appear from time to time but they are often caught, hauled ashore and cut up ready for market. To the east, the Sunda Shelf is thought to be a migration route between the north-west Pacific and the north-east Indian Ocean, which means white sharks have appeared off Malaysian Borneo and in the Lombok Strait. Migrants also come close to South Pacific islands such as Tonga and New Caledonia at certain times of the year when large baleen whales are visiting, especially humpback whales giving birth.

Around the coasts of China, white sharks are a protected species, with miscreants fined heavily or receiving a sentence of up to ten years in prison; indeed, there was a well-publicised case of a TikTok food blogger cooking white shark on camera and posting the video in January 2023. She was fined 125,000 yuan (US$18,500) and her website was closed down. Taiwan and Japan have had some very large white sharks recorded, such as the unconfirmed 7-metre (23-foot)-long specimen caught off Taiwan's Hualien County in May 1997; despite a ban on killing white sharks introduced by Taiwan in 2020, they are still caught illegally and sold in the region. Fins are sent to Hong Kong from Taiwan and as many as eighty other countries, it's thought, despite a ban on trade in white sharks and their parts. Also in the Pacific, Aniva Bay in the north-west, at the southern tip of Russia's Sakhalin Island, along with the Strait of Tartary, which joins the Sea of Okhotsk with the Sea of Japan, are the northernmost records of the species in the region. The Bohol Sea in the Philippines has been host to the southernmost.

Chile in the south-east Pacific was once a major centre, but overfishing has seen a big decline, although scientists discovered a 'fossil' white shark nursery here dating to Pliocene times (5.3 to 2.5 million years ago), testament to there being greater numbers in the past than today. Fossils of the ancestors of white sharks are also found here and in Peru and California, indicating that the white shark probably had its origins in the Pacific (see also Chapter 5). White sharks are still around, and one made itself unpopular when it decapitated a diver collecting shellfish near the resort town of Pichidangui, 150 kilometres (93 miles) north of Valparaiso in January 1980, while another attacked a teenage girl and a young male researcher snorkelling from a research ship about 300 kilometres (186 miles) east of Easter Island in March 1984.

On the other side of the continent, in the south-west Atlantic Ocean, large female white sharks up to 5.3 metres (17 feet) long have been caught off southern Brazil and elsewhere along the Brazilian coast, and there are records from Uruguay and

Argentina. It's thought that these sharks are part of large-scale migratory movements in the South Atlantic, which include the Tristan da Cunha archipelago, the remotest inhabited islands in the world. On other oceanic islands, such as St Helena and Ascension Island, white sharks are occasional visitors; indeed, white sharks have been tagged off St Helena, and Ascension was where two white sharks recently attacked a paddleboarder – he survived.

At the Cape Verde Islands in the North Atlantic, fishermen have landed exceptionally large specimens, the white sharks thought at one time to be attracted there by an open boat sperm whaling industry. They also occasionally visit the Canary Islands, turning up off Tenerife and La Gomera, but with only six shark bite incidents since the sixteenth century, the chances of being caught out are very low. Madeira also has the occasional visitor, but these visits are few and far between.

The Mediterranean Sea has been another hotspot, albeit nowadays with seriously declining numbers due to illegal fishing and bycatch. White sharks feature frequently in the writings of the ancient Greek and Roman naturalists and philosophers so, at one time in the distant past, there must have been a sizeable population (see Chapter 2). Elsewhere in the north-west sector of the North Atlantic, there have been a few sightings off western Europe, such as Portugal and the Bay of Biscay, but, alas, no confirmed sighting off the coasts of the British Isles (see also Chapter 6).

All this means, in fact, that white sharks can occur just about anywhere in the Earth's oceans, from subpolar waters to the tropics. They're absent from the most northerly parts of the Arctic Ocean, the most southerly parts of the Southern

Ocean and, curiously, the Arabian Gulf, where conditions are clearly not comfortable for white sharks. It makes the white shark the most widely distributed species of shark on earth.

It's also a remarkable but greatly misunderstood animal, something that maybe our biography can try to put right. Generally speaking, a biography is a detailed description of an individual's life, but, in this case, it's a comprehensive look into the lives of many individuals that lived from millions of years ago until the present day, giving rise to an ocean predator that evolution has honed almost to perfection, but also one that, down the years, has had its unfair share of bad press and could even be threatened with extinction in the not too distant future. For this reason, it would be good if we could have some kind of empathy with the main character, so somehow, we must tap into that delicate interface between humans and wild nature and show just how amazing is this animal – a fish, yes, but one with extraordinary qualities, a lord among sharks.

PART I

OF SHARKS
AND PEOPLE

CHAPTER 1

ANCIENT PEOPLE
OF THE SEA

In 1996, the bones of a young man were found in the On Your Knees cave, also known to First Nations as *Shuká Káa*, meaning 'Man Ahead of Us'. It lies at the northern tip of Alaska's Prince of Wales Island, where the man lived at the time folk were migrating from Asia to North America at the end of the last Ice Age. The exodus is thought to have started about 30,000 years ago, just prior to the Last Glacial Maximum, when a single population of people began to move eastwards through Siberia from what is now north-eastern China. By about 20,000 years ago, when climate change brought in milder and wetter weather, they were able to cross into the Americas because North America and Asia were joined by a land bridge, the result of sea levels being about 50 metres (165 feet) lower than they are today. This pioneering group was thought to have found an ice-free corridor along the continental divide between two adjacent ice sheets – the Cordilleran ice sheet next to the coast and the Laurentide ice sheet farther inland – and were able to travel into the heart of North America to the south of the Canadian ice sheets. Here, they had the entire land and all the game – woolly mammoth, steppe buffalo and other large mammals – to themselves. There was nobody else about. However, others soon followed.

Some arrived via the 'Kelp Highway' coastal route, and they probably came by sea rather than overland. The Laurentide ice sheet came right down to the edge of the ocean where the towering ice fronts of mighty glaciers collapsed into the sea, forming gigantic icebergs. The only way to follow the coast to the south, therefore, was by boat, probably in craft similar to the seal-skin umiak used by the Inuit to hunt bowhead whales. The young man in the cave was more than likely a descendant of those early maritime travellers.

Many of the coastal settlements in the Pacific Northwest, some close to 15,000 years old, were inundated when the ice sheets of the last Ice Age melted and sea levels rose, so only underwater archaeologists with scuba gear can explore them now, but the On Your Knees cave was on higher ground, which became an island, and so it is readily accessible; at least it became so when all the debris which had fallen down and blocked the entrance had been removed. Today, the cave is about a kilometre (0.6 miles) from the sea and 125 metres (410 feet) above sea level. It had been a refuge for a succession of bears, but the most interesting discovery was the human remains and the young man who once lived there, albeit for a relatively short time: he died when just twenty years old. The date was roughly 10,300 years before the present.

Tests on materials excavated from the cave indicated that the young man's diet was mainly marine mammals, in particular seals, and wherever there are seals there is sure to be a formidable competitor and serious danger – the white shark. You can imagine the scenario: young hunter sneaks up on a seal or sea lion rookery but inadvertently spooks the animals and they, almost as one, head for the sea. He takes to his canoe, a rather flimsy craft covered with animal skins or birch bark, and he manages to harpoon one in the water. After a struggle, there's blood everywhere. Cue white shark. It grabs the seal, and a tug of war is won by the shark. It must have been tough being a hunter here in those days, but he had to provide for his family. Ironically, his harpoon was probably tipped with a shark's tooth... but did white sharks really come this far north?

Given their ability to withstand cold temperatures, it may be no surprise to learn that in the Pacific Ocean, white sharks are often seen as far north as the Gulf of Alaska, and there is one record from the Bering Sea. It can live in such chilly waters because it has a special heat exchange system in its body, which you'll read about in Chapter 6. This enables the shark to maintain a body temperature slightly higher than the temperature of its surroundings, so it can function fairly normally in frigid waters. It has led scientists to realise that the white shark is present at higher latitudes more often than was previously thought.

A white shark, for example, lashed out at commercial fishing gear off the west coast of Vancouver Island, British Columbia, in the summer of 1961, and another individual was captured off south-east Alaska in 1981. All in all, twenty-nine reliable sightings or strandings were recorded in Alaskan and British Columbian waters between 1961 and 2004 in a survey by the late R. Aidan Martin, published in 2004. About 95 per cent of them were large fish, one giant with an estimated length of 6.2 metres (20 feet), which means that, while out hunting and fishing along the Pacific coast, those first Native Americans probably encountered white sharks, and big ones too; and there and then was started that uneasy relationship between sharks and people which *Jaws* encapsulated so graphically nearly ten thousand years later.

In Martin's survey, a good few great white sharks – twelve to be precise – were recorded around the islands of Haida Gwaii, just 75 kilometres (47 miles) to the south of Prince of Wales Island and part of British Columbia. This is home to the Haida people, and it has been so for at least 12,500 years, when the ice sheets were still in place and long before the Europeans came in 1774. They, like other tribes along the Pacific coast, are people of the sea, probably also descendants of the migrant pioneers, and the white shark is a creature they were in awe of, which they show in an interesting way.

Sharks, the white shark and dogfish in particular, often feature on the tall and spectacular crest and story poles carved by the Haida people (also known popularly but inaccurately as totem poles, the word 'totem' having been borrowed from *doodem*, meaning 'clan', as used by the Ojibwe people of the Great Lakes). At the Haida Heritage Centre, for example, there is a crest pole featuring four blue eyes and a mouth with red lips and triangular teeth that looks distinctly like a white shark with its jaws extended at the moment of an attack. Maybe one is reading too much into this carving, but it's there for all to see, as was a large white shark that was washed up on a beach on Graham Island, the largest island in Haida Gwaii group, just as they must have been 12,000 years ago. The stranding occurred in December 1977, and the shark was 5.5 metres (18 feet) long. The species is not common here but, from time to time, it does make its presence felt. Photographs exist of a dead porpoise with white shark bites and a Steller sea lion with horrible injuries, but still alive, that were taken on Haida Gwaii beaches. They represent a fish that was sure to impress.

In fact, sharks must have played an important role in Haida life to feature so strongly on crest and story poles, along with bears, eagles and the other animals they would have encountered and borrowed to represent their particular clan. The traditional shark motif – triangular teeth and thick red lips – is picked up again in a modern rendering on a crest pole erected in the summer of 2022 in front of their longhouse in Old Masset by renowned Haida artist Kihlyaahda Christian White and his wife Candace Weir-White. The totem of red cedar is 16 metres (53 feet) tall and 1.5 metres (5 feet) wide at the base and each of the figures, carved into the wood one above the other, is based on a traditional story. The shark, for example, is a white shark mother who captured a woman, and that woman became a shark woman. Similarly, there is the story of the dogfish mother, in which the shark represented is the smaller dogfish, the spur dogfish by all accounts, as the prominent spine at the front of its dorsal fins is found in ancient middens. A woman, so the story goes, was on a rock on the shore of Haida Gwaii, and the dogfish that lived there led her down to the seabed, where it removed its dogfish blanket to reveal that it was actually human. The woman grew fins on her arms, legs and head and she never went back to the human world on the shore ever again, but her clan's descendants have used the dogfish image on its crest pole ever since. Stories like these, in which people and animals are transformed, have been handed down from one generation to the next for millennia, and the Haida keep them alive to this day.

Vancouver Island, especially the west coast, is another known white shark location. Positioned about 250 kilometres (155 miles) to the south-east of Haida Gwaii, it is home to the Nuu-chah-nulth people. They were humpback whale hunters for more than 4,000 years, and wherever there are dead whales, there are sure to be white sharks. The large, mature sharks are attracted to floating carcasses, so the hunters must have encountered and had great respect for them, especially the monstrous ones that they say 'liked to eat canoes'. They were named 'dogfish mothers', the name borrowed from the spur dogfish, and during a ceremonial feast, a dancer would appear, wearing the *mamachaktl hinkiitsim* or shark rider headdress, and then sway this way and that as if swimming underwater on the back of a gigantic shark. The Nuu-chah-nulth people were also among the first conservationists; in fact, they have a word for it – *uhmuuwashit*, which means 'keep some and not take all', an enlightened attitude to their fishery. As Chief Earl Maquinna George of the Ahousaht First Nation once said, 'these people knew how much they could use, and they didn't ever use it all.'

On the other side of the country, on the Atlantic coast of Canada, the Mi'kmaq are a First Nation people that live in Nova Scotia, New Brunswick, Prince Edward Island and Newfoundland, and they have been there for at least 4,000 years. They have a deep relationship with nature and throughout their history, they have been confronted with the great white shark; they have a specific name for it, *mukumu'k*, and another for sharks in general, *wipitamekw*, meaning 'many-toothed one'.

Today, large white sharks are present as far north as the

Gulf of St Lawrence and Newfoundland, arriving mainly in summer and autumn, and there is no reason to believe there were fewer of them thousands of years ago. Archaeologists have found sharks' teeth in shell middens dated 4,000 years before the present. The teeth had holes drilled in them, so they were likely worn as pendants, which means these ancient people must have had a significant relationship with the white shark: probably a sense of awe mixed with fear. However, the relationship stopped abruptly when Europeans arrived: no more shark tooth pendants. What happened there, nobody knows.

Whether any of these First Nation peoples were the victims of shark bite incidents is rarely recorded, but a letter from Father Pierre Maillard, dated about 1755 and collected by Earle Lockerby, recalls a conversation with Arguimaut, a Mi'kmaq shaman on Prince Edward Island (or Ile Saint-Jean as it was known at that time). It tells of life before colonisation, in particular, the relationship with white sharks. It recounts how Mi'kmaq fishermen in their moose-skin-covered canoes were faced with attacks from a 'bad fish'. 'All too often,' the shaman recalled, 'these malicious beings attacked the sterns of our canoes, so suddenly and without warning they sink the boat and all who are in it. Some of the crew escape by swimming, but there are always some who fall prey to these voracious flesh-eating fish.'

The description of canoe attacks was similar to those of the Nuu-chah-nulth people on the west coast, except that the Mi'kmaq apparently fought back. By way of defending themselves, the fishermen tried to harpoon the shark and fend it off until they could reach land, but there was also a more ingenious defence. They tried to trick the shark into thinking it was closer to land than it really was and so there was the real risk of it beaching. They fastened foliage to the stern of the canoe, split spruce roots and tied them to the bottom of the canoe with eel grass, and decorated it with any other plants that were available. To the shark, it looked like it was too close to shore.

Whether these 'bad fish' are great white sharks is called into doubt by some commentators. They believe the miscreants were orcas or killer whales, common enough on both the east and west coasts of North America. However, there are no records in historical times of orcas killing people, although a recent spate of attacks on yachts, especially their rudders, in the Strait of Gibraltar gives credibility to these creatures attacking the stern of canoes. There is, though, another reference from another priest writing in 1824.

Father Vincent de Paul was travelling by canoe from Tracadie, Nova Scotia, to Cape Breton Island, which would have taken him past Prince Edward Island, when his paddlers, probably from another clan, 'perceived three monstrous fish called *maraches*'. 'They were frightened,' recalled the priest, 'as these fish are very dangerous. Their teeth are like gardeners' knives for cutting and boring, or like razors slightly bent. They are extremely voracious, and often follow boats, attacking them with violence.'

The description given is certainly not of three orcas, as they have fairly blunt, conical teeth. The 'razors slightly bent' could only refer to one thing – our old friend the white shark, and recent behavioural research in the eastern Pacific has revealed that it often hunts with a friend, not cooperating, but one following the other and taking advantage of any successes (see also Chapter 8). The three *maraches* could well have been white sharks, the word *marache* thought to be from the Basque name for 'shark'. They were considered so dangerous because their teeth munched through the flimsy, bark-covered canoes like a knife through butter. The craft became waterlogged, and their crews left floundering in the water – easy pickings.

More recently, in 2021 to be precise, a 21-year-old woman was swimming in the same area, off the coast at Cape Breton, when she was attacked by a white shark. According to the Canadian Shark Attack registry it was the first shark bite incident since the 1800s, although scientists predict there are likely to be more in the future. The reason is that the white shark population is recovering from the seal-hunting days, when the population of their primary prey – mainly harbour and grey seals – was decimated. With the seals gone, the sharks didn't bother to visit, and so were rarely seen. When the seals were protected, however, their population increased significantly, and the sharks returned. It's possible we could be seeing in the not-too-distant future what it was like all those years ago, when there were many more white sharks in the area than there are now.

Although there are few reports of shark bite incidents back then, for obvious reasons, it seems to be beyond doubt that First Nation peoples were harassed and even lost their lives to white sharks, but there is one record from Japan that's unequivocal – there's a body! One of the oldest shark bite victims has been uncovered in an ancient Jōmon culture site – the Tsukumo Shell Mound – in the southern part of the island of Honshu. Alongside the shell middens in the village cemetery were the gruesome remains of an adult male, estimated to be close to 3,000 years old. The body, or what was left of it, had 790 perimortem traumatic lesions characteristic of a serious shark attack. Researchers at Oxford University found that the bite marks were mainly on the arms, legs, pelvis and ribs and they were able to build a three-dimensional model of the victim's bones and injuries. It revealed that he had probably tried to fend off the shark with his left hand, but the shark took it. A later bite severed arteries in the leg and the man succumbed through loss of blood. The shark made off with his right leg. The left leg became detached and was found in the grave resting on his chest. The culprit was more than likely a white shark, the species responsible for modern shark bite incidents in the same area, although tiger sharks are present too.

If the man was buried close to where he was killed, the incident must have taken place in the vicinity of the Seto Inland Sea, where there have been several attacks by particularly large white sharks in modern times, including a professional shell diver who was killed by what was identified as a white shark as recently as March 1992. The general belief was that these were isolated events, with few white sharks visiting Japanese waters, but a survey by Kazuhiro Nakaya published in 1994 found that they are much more common than was first thought. During the eighteen months from March 1992 to August 1993, he discovered there were at least nineteen white shark captures or observations reported in Japanese waters with the likelihood that many more captures during the same period have gone unrecorded. Four were in the Seto Inland Sea, one specimen over 5 metres (16 feet) long, a large shark. However, it's possible that they could have been even bigger all those years ago, before the animal was hunted intensely; as sharks can grow old, so they could grow *very* big.

However, what is probably the oldest known shark attack victim was discovered near the Peruvian town of Paloma, where the bones of a seventeen-year-old boy were estimated to be 6,000 years old. Like the Japanese skeleton, it showed signs of a shark attack. Hip and arm bones had deep bite marks, and the left leg was missing. The boy's grave, however, was unusual. According to researchers from Harvard University and the University of Missouri, who excavated the site, most graves were covered with soil or located below a house floor and would have been close to or under the village's reed huts, but this grave was a basic open pit covered by a grid of canes overlaid with woven mats to form a roof. A seashell, a large flat rock and several pieces of rope, some tied with elaborate knots and one with a tassel, had been placed alongside the body.

The site is about 3.5 kilometres (2 miles) from Peru's Pacific coast, where the Humboldt Current, flowing northwards from the Antarctic, is very rich in wildlife. When the wind blows parallel to the shore, upwellings drag up nutrients from the deep-sea floor which feed first phytoplankton, then zooplankton, followed by huge shoals of small fish, then large fish, fur seals and South American sea lions, until the energy stream reaches the apex predators, including white sharks.

In ancient times, the sharks certainly impressed the locals. They were revered, possibly even worshipped. One archaeological site where this is thought to have occurred is Huaca Pucllana, an ancient settlement on the central coast of Peru. It was occupied by people of the Lima culture and dated between 1,824 and 1,325 years old. The inhabitants were farmers and fishermen who built truncated pyramids made of clay bricks to worship deities symbolised by figures associated with the sea, such as waves, sea lions and, of course, sharks. Their mortal remains have been found in buildings where ritual banquets were held, with the teeth of white sharks common around the site, often in small piles where jars had once stood, so they must have had some significance in proceedings, including gruesome human sacrifices. There are many ceramic jars at the site, the artwork depicting many aspects of life at Huaca Pucllana. Some show the advanced reed boats they used for fishing, which could be up to 3 metres (10 feet) long, and one ceramic has a fisherman in a boat being attacked by a shark, no doubt a white shark.

About 100 kilometres (60 miles) to the south of Huaca Pucllana is another archaeological site at the town of Asia, on the coast. It too has white shark teeth alongside the remains of other marine life. And at Potrero Tenorio in central Peru, fragments of pottery show the distinctive shapes of white sharks.

It all means that in the distant past, white sharks must have been numerous along this stretch of coast in order that there can be so many teeth at the various archaeological sites; but, like the situation in eastern Canada, the hunting of sea lions and fur seals to the edge of extinction caused Peru's white sharks to disappear too. The seals have bounced back, but the sharks are still relatively rare. Only two reports exist in the scientific record, and they are both old, from the 1940s. In 1944, pictures of a 5-metre (16-foot)-long white shark, which was caught in a net outside Ancon Bay, to the north of Lima, appeared on the front page of the *Andean Air Mail and Peruvian Times* and in 1949, another shark, just under 5 metres, was reported at Miraflores, a district in Lima province. There is a record of a fatal shark attack in Peru in 1849 and fatal white shark attacks are known from Totoralillo in 1963 and El Panul in 1980, both in neighbouring Chile, and that's about it.

However, on the other side of the continent, on the Atlantic coast of Argentina, two teeth from a great white shark turned up in an unexpected place. This time the archaeological site – the Nutria Mansa locality in the province of Buenos Aires – was that of land-based hunters, the hunter-gatherers of the Pampas. They lived in the early-middle Holocene, about 8,000 years ago, and more usually hunted terrestrial mammals, especially guanacos, rather than sharks, so what were sharks' teeth doing among their possessions?

The teeth had an artificial transverse groove at the root tip, maybe for a thread to be tied. It's an indication, perhaps, that they were adornments, possibly pendants, although blunted serrations suggest they were also used as tools, probably for cutting. They could have been worn around the neck a bit like somebody today with a penknife in their pocket. How these people obtained the teeth, however, is unknown. They could have chanced upon a beached shark or come across a dead seal which had been attacked by a shark, escaped, but had the teeth embedded in its fatally injured body. The teeth could have been traded, or they simply found the teeth washed up on the shore. Whatever the reason, it was not a lone case. White shark teeth have also been found associated with terrestrial hunters in Uruguay and Brazil.

In Uruguay, shark teeth, along with seal teeth, were found at an archaeological site on San Miguel Hill, about 30 kilometres (19 miles) from the coast. They were in a ritualistic setting associated with the mound builders of the Cerritos and dated to between 2,500 years ago and historical times when written records began. Another site at the coast, Punta de la Coronilla, dated 2,700 years before the present, has white shark teeth, along with seal and fish bones. It's thought the two sites might be linked and that the people had a somewhat complex lifestyle, spending some of their time inland and the rest at the coast.

In Brazil, the focus of attention has been the sambaquis or shell mounds that are found along the Atlantic coast from the border with Uruguay to just north of Rio de Janeiro, the mounds decreasing in size and frequency the farther north you go. They are basically ancient waste tips of people who hunted and fished and lived between 6,000 and 2,000 years ago. The middens contain sharks' teeth and the remains of both marine and terrestrial animals. While many of the teeth have simply been discarded, some have perforations and were probably worn as pendants, while others made up necklaces. However, there was more to it than that.

Researchers at the Rio do Meio archaeological site in Florianópolis, a so-called 'shallow site' rather than a mound, took a close look at tooth microwear and so were able to work out the function of sharks' teeth in this precolonial society. While many were ornaments, as we have seen, this research indicates that the teeth were also most definitely used as tools. The work was so detailed that it could reveal whether the tooth was used primarily to cut soft materials, such as leather and meat, or hard stuff, such as bone and wood, and whether it was used for piercing, cutting or scraping. One great white shark tooth even showed that it was employed to cut away bark from green wood. Hafted sharks' teeth also made effective arrowheads. They helped the arrow to penetrate the prey's body, the only drawback being that they shattered if they hit bone. On the plus side, they floated in water, while arrows tipped with stone sank, so the shark-tooth-tipped arrows could be recovered and used again. All in all, it is a remarkable piece of work, literally at the cutting edge of archaeological research, and it highlights the special relationship that must have existed between humans and sharks along the Brazilian coast in precolonial times.

In fact, sharks' teeth are so numerous at the many archaeological sites on the Atlantic coast of South America that there must have been a healthy population of white sharks here too, but today it's a shadow of its former self. A survey of the scientific literature published in 2001 came up with just nineteen records of white sharks in Brazilian waters, along with one offshore sighting, and another paper quoted 'about 24', so these days the fish is not common. However, a historical record from 1992 revealed one of these rare giants, which was caught accidentally in a fisherman's gillnet off Bom Abrigo Island, about 55 kilometres (34 miles) off the coast of southern Brazil, and it was quite an animal; in fact, it was so big that two fishing boats were needed to tow it back to port. It was a large female, 5.3 metres (17 feet) long, with an estimated body weight of 2.5 tonnes (5,500 pounds), or about one-and-a-half times the weight of a family car. When her stomach was opened up several shark heads spilled out – two sandbars, two scalloped hammerheads and a blue shark, most likely discarded from fishing boats – and the remains of one bony fish and two dolphins, one of which was a mature Atlantic spotted dolphin. The remarkable thing about her, though, was the size of her liver. The hepatic somatic index, the ratio of liver weight to body weight, was 27 per cent, which makes it one of the largest white shark livers in the scientific record.

The female was living where seasonal upwellings bring cold deep-sea waters into the coastal environment, at a point where the continental shelf is narrow, an area of coast that encompasses the shoreline between the states of Espírito Santo and Rio de Janeiro. Three other large and mature white sharks – 5.0 metres (16 feet), 5.3 metres (17 feet) and 5.5 metres (18 feet) long – were caught in the Ceará region, which experiences similar conditions to those where the large female had been caught. So, even though white sharks are few and far between in these waters today, those that have been observed have made a big impression on the scientific community. If sharks of such size had been around when ancient tribes had been diving, say, for shellfish and other foods from the sea – and the evidence seems to suggest that they were – then those folk must have been firmly in the danger zone, so it's little wonder that they revered such a large and powerful fish.

CHAPTER 2

THE SELACHIA

Fragments of a ceramic vase, the Cretere del Naufragio, dated 725 BCE, were excavated at Lacco Ameno on the Italian island of Ischia in the Bay of Naples. On its sides are depicted shipwrecked sailors fighting to stay alive, and they're surrounded by myriad sea creatures. One person has his head in the mouth of a huge fish, probably a white shark. It is Italy's oldest example of figurative vase painting and the world's first report of a shark attack.

This association between sharks and people in the Mediterranean has a history that started way before the eighth century BCE. It was probably gustatory at first, as the remains of sharks have been found in late Stone Age caves in southern Italy and Spain; but wherever sharks, especially white sharks, and people come into contact there is always the prospect of the prey turning into the predator.

Centuries later that love–hate relationship, which underlines the story of *Jaws*, engaged ancient Greek and Roman poets, philosophers and naturalists. Take Herodotus (*c.*484–*c.*425 BCE), for example, the 'father of history', at least according to Cicero. In his extensive tome *Histories*, he tells of a time, in 492 BCE, when the Persian fleet were getting the upper hand over the Greeks. However, their luck suddenly changed when they were struck by a powerful north wind, and their ships were dashed against the cliffs of Mount Athos. An estimated 300 vessels were lost, and their crews tossed into the sea. Waiting for them were sharks – among them probably white sharks. Attracted by the pandemonium, they moved in and many of the 20,000 Persians in the water perished, either by drowning or consumed by the sharks, which Herodotus describes as 'sea monsters'.

Aside from the sheer numbers of people involved, and I'm sure there was some degree of exaggeration, the story gives an indication of the abundance of sharks that must have lived in the Mediterranean Sea at that time, a far cry from the few that are seen today. Even so, some recent catches give a validity to Herodotus's sea monsters, as some of the world's biggest white sharks have appeared in the Mediterranean. Off the south-west coast of Mallorca in the 1920s, for example, two enormous white sharks were caught in tuna traps – almadrabas, the world's oldest form of industrial fishing. Both were reported to be a staggering 7 metres (23 feet) long. If the measurements are correct, they would be the largest white sharks ever to have been recorded but, alas, we'll never know whether their lengths were inflated.

Even so, there have been exceptionally large white sharks caught in the Mediterranean, and their presence was sometimes a bit of a shock. One enormous individual captured in 1987, for instance, had a blue shark, a large swordfish and a dolphin in its stomach. In earlier times, sharks might have had much more sinister stomach contents. The poet Leonidas of Tarentum, who lived during the third century BCE in what is now Puglia on the heel of Italy, tells the shocking tale of the sponge diver Tharsys, son of Charmides. Tharsys dived down to loosen an anchor and just as he reached out to be pulled back into his boat a huge 'sea monster', most probably a white shark, grabbed him and bit him in two, swallowing the lower half of his body. The rest of him was dragged on board and later buried on the beach, thus poor Tharsys was buried 'both on land and at sea'.

In the previous century, a great scholar who had an appreciation of the fishes living in the Mediterranean and recognised which were potentially dangerous to people was none other than the ancient Greek philosopher and naturalist Aristotle (384–322 BCE), the generally agreed 'father of zoology'. His studies on sharks, which he called 'selachia', are groundbreaking. Whether these observations were made by Aristotle directly or collected by his acolytes we'll never know, but however they were obtained, they were very perceptive and amazingly accurate.

Take, for instance, a basic anatomical reveal. Aristotle pointed out that one way to tell the sexes apart was by the presence or absence of claspers. Claspers are like a pair of permanently pumped-up penises that dangle under the body close to the vent. They also have a similar function, that of delivering sperm, so male sharks have them, females don't. He further noted that sharks and rays have 'uncovered gills' and they are 'gristle-spined', that is, they have a cartilage skeleton; and he worked out that some types of sharks, rather than depositing eggs, give birth to 'live' young, writing: 'Selachia and vipers, though they bring forth their young alive externally, first of all produce eggs internally.' Aristotle went on to reveal that the shark embryos are first nourished by yolk from a yolk sac, and when that's used up, a placenta forms between the embryo and the mother, and nutrients come directly from her until it is time to give birth. And where she would have given birth was another revelation.

No doubt, local fishermen were good sources of information. In those days, with white sharks more common than they are today, fishermen must have come into contact with them almost daily. They would have known where the large, mature and possibly pregnant females and the newborn youngsters were to be caught, and so Aristotle was the first scientist to put forward the notion of 'shark nurseries', something we take for granted nowadays, but then, in the fourth century BCE, he introduced it as a completely new idea. He wrote: 'The selachia come in from the high seas and out of the deep water towards land and produce their young there; this is for the sake of the warmth and because they are concerned for the safety of their young.' It means that Aristotle, or the fishermen he consulted, observed adult white sharks and young-of-the-year, and put two and two together… and not a tag in sight.

And there the matter of shark nurseries rested for more than two thousand years until, in 1916, Professor Alexander Meek authored *The Migrations of Fish*, in which he mentions that female dogfish migrate to a particular region for the liberation of their young. Aristotle was certainly way ahead of his time.

The Roman philosophers and naturalists that followed Aristotle, such as Pliny the Elder (23/24–79 CE), who wrote the encyclopaedic *Natural History*, which became a model for all the encyclopaedias that followed, and Claudius Aelianus (175–235), who authored *On the Nature of Animals*, a mixture of facts and fables, were more compilers of information than hands-on naturalists. They, and authors like them at the time, were responsible for keeping the older Greek texts alive, so nothing was lost. They did, however, add a few things to the body of knowledge.

Pliny, for example, was one of the first to document scientifically the interactions between sharks and people. His focus was on sponge divers. 'Divers have fierce fights with the dogfish,' Pliny wrote, 'dogfish' referring to any type of shark, large or small, but the harassment of sponge divers he describes has the hallmarks of attacks by white sharks.

'These attack their loins and heels and all the white parts of the body,' he continued.

The one safety lies in going for them and frightening them by taking the offensive; for a dogfish is as much afraid of a man as a man is of it, and so they are on equal terms in deep water. When they come to the surface, then the man is in critical danger, as the policy of taking the offensive is not available while he is trying to get out of the water, and his only safety is in his comrades. These haul on the rope tied to his shoulders; this, as he carries on the duel, he shakes with his left hand to give a signal of danger, while his right hand grasps his dagger and is occupied in fighting. Most of the time they haul gently, but when he gets near the boat, unless with a quick heave they suddenly snatch him out of the water, they have to look on while he is made away with. And often when divers have already begun to be hauled up, they are snatched out of their comrades' hands, unless they have themselves supplemented the aid of those hauling by curling up into a ball. Others of the crew of course thrust out harpoons, but the vast beast is crafty enough to go under the vessel and so carry on the battle in safety.

'Consequently,' according to Pliny, 'divers devote their whole attention to keeping a watch against this disaster; the most reliable token of safety being to have seen some kind of flat-fish, which are never found where these noxious creatures are – on account of which divers call them the holy fish.'

This last reference to the 'holy fish' picks up on something Aristotle referred to more than 400 years earlier in his *History of Animals*. 'Wherever an anthias-fish is seen,' he wrote, 'there will be no dangerous creatures in the vicinity, and sponge divers will dive in security, and they call these signal-fishes "holy fish"'. Today, the *Anthias* mentioned by Aristotle refers to a genus of fish from the grouper family and not a flatfish at all, but Pliny refers specifically to a flatfish as the holy fish, and, lo and behold, there is actually a flatfish known to deter sharks. It's called the Moses sole, and it exudes a milky ichthyotoxin from the base of its dorsal and anal fins that repels sharks. It could be the identity of the 'holy fish', but it is not present in the Mediterranean Sea, only in the nearby Red Sea and western Indian Ocean, which, I guess, is still in Pliny's geographical sphere of interest; so, it begs the question: does this mean sponge divers in the Red Sea knew about the deterrent nature of the Moses sole all those years ago, in 77 CE or thereabouts, something that was not published in the scientific literature – by Eugenie Clark and Anita George of the University of Maryland – until as recently as 1979?

Before we leave the ancient world, maybe extracts from a Greek didactic poem would be uplifting. The writer is the

Greco-Roman poet Oppian, who lived in the second century CE. The work, written in about 177–80, is dedicated to the emperor Marcus Aurelius and his son Commodus – the very same, for *Gladiator* fans – and the subject is fish and fishing, all five volumes of it, but one book focuses on a hunt for a white shark. After a longish introduction, Oppian gets down to the nitty-gritty.

For these monsters the line is fashioned of many strands of well-woven cord, as thick as the forestay of a ship … The well-wrought hook is rough and sharp with barbs projecting alternately on either side … A coiled chain is cast about the butt of the dark hook – a stout chain of beaten bronze to withstand the deadly violence of his teeth and the spears of his mouth … For fatal banquet they put upon hook a portion of the black liver of a bull or a bull's shoulder suited to the jaws of the banqueter. To accompany the hunters, as it were for war, are sharpened many strong harpoons and stout tridents and bills and axes of heavy blade and other such weapons as are forged upon the noisy anvil…

And when he espies the grievous banquet, he springs and disregards it not, obedient to his shameless belly, and rushing upon the hooked death he seizes it; and immediately the whetted hook enters within his wide throat and he is impaled upon the barbs. Then, roused by the wound, first, indignant, he shakes his deadly jaw against them and strives to break the brazen cord; but his labour is in vain. Then, next, in the anguish of fiery pain he dives swiftly into the nether gulfs of the sea. And speedily the fishers allow him all the length of the line; for there is not in men strength enough to pull him up and to overcome the heavy monster against his will ... Straightway as he dives they let go with him into the water large skins filled with human breath and fastened to the line. And he, in the agony of his pain, heeds not the hides but lightly drags them down, all unwilling and fain for the surface of the foamy sea...

Now when the deadly beast is tired with his struggles and drunk with pain and his fierce heart is bent with weariness and the balance of hateful doom inclines, then first of all a skin comes to the surface, announcing the issue of victory and greatly uplifts the hearts of the fishers ... And immediately other skins rise up and emerge from the sea, dragging in their train the huge monster, and the deadly beast is hauled up all unwillingly, distraught in spirit with labour and wounds...

Then the courage of the fishers is roused and with hasting blades they run their well-oared boats near ... Then one brandishes in his hands the long-barbed trident, another the sharp-pointed lance, others carry the well-bent bill, another wields the two-edged axe. All toil, the hands of all are armed with mighty blade of iron, and close at hand they smite and wound the beast with sweeping blows ... and all the sea is stained with the gory filth poured forth by his deadly wounds. The infinite water boils with the blood of the beast and the grey sea is reddened...

But when, overcome by the pains of many gashes, fate brings him at last to the gates of dismal death, then they take him in tow and joyfully haul him to the land; and he is dragged all unwilling, pierced with many barbs as with nails and nodding as if heavy with wine in the issue of deadly doom...

Even when he is killed and laid upon the land one still dreads to approach his corpse of dread aspect and fears him when he is no more, shuddering even when he is gone at the mere teeth in his jaws. At last, they take courage and gather about him in a body, gazing in astonishment at the ruins of the savage beast.

The poem is remarkable for many reasons, not least that it gives a vivid picture of the catching of a white shark in ancient Roman times yet makes the whole gory experience sound almost 'romantic' and, indeed, 'heroic'. It also describes in graphic detail events that were enacted in the fictional world of *Jaws*, including air-filled skins instead of yellow barrels when Quint, Brody and Hooper went out to catch the monster shark. It seems things haven't changed all that much in 1,900 years.

After the decline and fall of Rome in the fifth century CE, there followed nearly a thousand years of what was purported to be a time of little scientific or cultural endeavour. It was the so-called Dark Ages, and they were supposed to have lasted until the fifteenth century, although many historians today don't think the Dark Ages in Europe were that dark. The curmudgeon who threw a wet blanket over any scientific or cultural undertaking was the Italian poet Petrarch, who, by all accounts, appears to have had a gripe with his fellow poets in Italy during the 1300s. Unfortunately, his angst was catching, and every day became a dark day, but the reality was things scientific and cultural progressed almost as normal; after all, the printing press was invented during those dark days, and that's no mean achievement.

There were also bestiaries, beautifully illustrated books that interpreted what Europeans knew about plants and animals, real and imagined. Sharks appeared, but you'd be hard-pressed to recognise one. There were, for example, several illustrations of a creature with a fish-like body, the head of a dog and fins like feet and legs, which were labelled 'sea-dog', otherwise known as canicula, and given the Latin name *Canis maris*, although today, *Canis* is the genus name for members of the dog family.

Such a 'dog of the sea' appears in a twenty-volume encyclopaedia by Thomas of Cantimpré (*c*.1201–*c*.1272), a Flemish friar and writer, preacher and theologian. Over nearly twenty years, from 1225 to 1244, he wrote *Liber de natura rerum*, which roughly translated means 'The Nature of Things'. Book 6, with the title 'Marine Monsters', includes *Canis marinus* or the sea-dog. Thomas describes it thus:

> The dogs of the sea are marine animals, as Pliny once wrote, terrible in ferocity. They have the strongest teeth shaped like nails. They hunt the shoals of fish through the sea like wild dogs on land, except that they cannot bark, but instead have a terrible breath. The beast can be killed but with difficulty by many tridents.

What type of shark Thomas describes is anyone's guess, and his description is typical of the time. During the medieval period anything vaguely scientific about sharks is notable by its absence. Things really didn't get going again until the supposed Dark Ages were little more than a distant memory and the Italian Renaissance spurred everyone on.

First up was French naturalist Pierre Belon (1517–64), who took 110 different types of fish, including a few sharks, and tried to draw up a comparative analysis of them. It was, in fact, one of the earliest attempts to classify the white shark in the scientific literature and was published in Belon's 1553 tome with the tongue-twisting title *De aquatilibus duo, cum eiconibus ad vivam ipsorum effigiem quoad ejus fieri potuit, ad amplissimum cardinalem Castilioneum*. In the scientific work, whose title refers to the making of life-like effigies, there is a realistic woodcut of a white shark, which he gave the scientific name *Canis carcharias*, so it's thought Belon tagged it to the 'sea dog'. For his work comparing the anatomical features of fishes, among other creatures, Belon was acknowledged as the founder of the scientific discipline of comparative anatomy.

Belon was one of three scientists who were born about the same time and who moved away from the fanciful animals of the bestiaries and to a more truthful and carefully record-ed view of the natural world. The others were the Italian naturalist and geologist Ippolito Salviani (1514–72) and the French naturalist Guillaume Rondelet (1507–66). In his 1554 book *Libri de Piscibus Marinus in quibus verae piscium effigies expressae sunt*, Rondelet called the white shark 'De Lamia', a child-eating fiend in Greek mythology and a name still used in Greece today, while the similar-sounding 'lamie' is used in southern France; in fact, during Rondelet's time, the white shark or lamie was relatively common along the Mediterra-nean coast of France, especially the 350-kilometre (220-mile)-long stretch between Sète and Nice. Here a white shark was hauled ashore, and its stomach cut open to view the contents. To everyone's surprise, out spilled a knight's complete suit of armour. Presumably its owner had already been digested.

No doubt Rondelet came across many white sharks, and big

ones too, quite capable of swallowing a knight whole. And, just to show what the white shark is capable of swallowing, a more recent report, and a gruesome one at that, was published in 1909 by two Italian zoologists, Dr M. Condorelli and Dr G. Perrando. They describe how a female white shark, 4.5 metres (15 feet) long – not huge for a white shark – was caught in a net off Capo Santa Croce, eastern Sicily, and, when it was cut open, they discovered the intact bodies of three people – two adults and a child – plus the remains of a cow and a dog inside. It is thought they had all drowned as a result of the December 1908 Messina earthquake and overwhelming tsunami and the shark had scavenged the bodies.

So, even relatively modest-sized sharks like this can swallow a person whole. This kind of revelation caused Rondelet to question the Biblical story of Jonah and the whale: for 'whale' read 'white shark'. He wrote: 'It is probable that this was the fish of Jonah rather than a whale. A whale does not have a throat wide enough to swallow a man whole and later regurgitate him.' In this, Rondelet was correct. The large baleen whales have very narrow throats, no more than a few centimetres across, about the size of a human fist, and if something large should get stuck in their throat they could easily die. They also have an oral plug, a fatty and muscular sac that swings upwards and blocks any water entering the lower respiratory tract when filter feeding, so the chances of Jonah being swallowed by a whale are zero. The white shark is a better bet. And there was an event that gives some credibility to Rondelet's hypothesis... that is, if it's true.

In 1758, according to Edward Bouverie Pusey's commentary on *The Minor Prophets* (1886), a sailor fell overboard from a frigate that was sailing in the Mediterranean in stormy weather. A monster white shark took the unfortunate seaman into his wide throat and swallowed him whole. His fellow crewmen leaped into the ship's sloop and tried to help their comrade. The captain of the ship ordered a small cannon to be fired at the giant fish and it was struck by the cannonball. It promptly vomited up the sailor, who was still alive and 'little injured', or so the story goes. The sloop picked him up and the shark was harpooned. It was estimated to be over 20 feet (6 metres) long and weighed 3,924 pounds (1,326 kilograms). It was dried and presented by the captain to the sailor. He then toured Europe, it is said, exhibiting both the shark and himself.

However, a slight digression from our white shark narrative might throw some light on stories like the above. Although baleen whales couldn't swallow you whole and regurgitate you later, it's possible a sperm whale could; after all, it usually slurps down the squid it catches whole, including giant and colossal squid, which are pretty big. And there is a nautical tale that recounts a human being swallowed by a whale. It goes like this:

It was the early 1900s off the Falkland Islands in the South Atlantic. A whaling ship had chased a sperm whale for several hours, with several sailors ending up in the water, but then finally it succumbed and was drawn up alongside the mother ship. The men, armed with flensing knives, sliced away the blubber and meat, working all day and part of the night. By the following afternoon, they were able to hoist the stomach, which appeared to have something moving inside, onto the deck. When they cut into it, there was one of the missing sailors doubled up and unconscious. He was gently laid out on the deck and doused in seawater, which began to revive him. However, his mind was not right, and he was placed in the captain's quarters, apparently a raving lunatic. The captain and officers gathered round and, over a period of three weeks, managed to bring him round and, having recovered from the shock, he resumed his duties.

Real, imagined or just a nautical fairy tale? Science tells us that methane in the sperm whale's stomach should have killed the sailor there and then, so the story looks to be an old wives' tale. It does, however, have a remarkable similarity to the white shark story. Maybe sea monsters swallowing sailors, followed by crew mates bringing the unfortunate deckhand back to life, is a universal seamen's tale told at harbourside inns the world over. A wonderful story, but the science is flawed, alas.

Rondelet, meanwhile, published his epic *Libri de Piscibus Marinis* or *Book of Marine Fish*, containing the description and illustrations of almost 250 sea creatures, including seals, whales and crustaceans, all of which were considered in those days to be 'fish'. There were also a few oddballs, like the monkfish and the bishop-fish, which really did look like a human monk and bishop, at least in the drawings. It was translated into French and republished in 1558 with the title *L'Histoire entière des poissons* or *The Complete History of Fish*, and it became the seminal work for budding ichthyologists (fish biologists) for many years afterwards.

Rondelet, Salviani and Belon were in the vanguard of sixteenth-century naturalists who brought comparative shark anatomy to life. They were followed by the likes of Swiss naturalist Conrad Gessner (1516–65), whose book *Icones Animalium* has a bizarre illustration of a monstrous fish – the

Ziphius, presumably a white shark as it has a seal in its mouth. He followed that with a rather dried and desiccated white shark illustration in his 1558 tome *Historia Animalium* (second edition). Next up was the Danish scientist Nicolaus Steno (1638–86), whose drawing of a dissection of a white shark's head is legendary, if not a bit scary.

However, there was more. Steno, together with Gessner and Italian naturalist Fabio Colonna (1567–1640), became embroiled in a growing debate on fossils. They were not understood at all, but Gessner found close similarities between fossil sea urchins and living ones and suggested that fossils are the petrified remains of living organisms, while, in 1616, Colonna put forward the notion that *glossopetrae*, so-called tongue-stones, were actually sharks' teeth. Steno agreed (see also the beginning of Chapter 5). Even so, how could fossils like these be found at the tops of mountains or embedded in rocks some distance from the sea? He wrote: 'Snails, shells, oysters and fish, found petrified on places far remote from the sea. Either they have remained there after an ancient flood or because the bed of the seas has slowly been changed.'

What Steno had realised was that the rocks were once sediments, such as mud, that had accumulated at the bottom of a now vanished sea or on its shores, and dead animals had sunk into the mud. More mud settled on top and, due to some geochemical process, the mud and the fossils turned to stone. Finally, major earth movements had taken them all somewhere else, even to the tops of high mountains. What clinched it for him was that fossils are only found in sedimentary rocks and not in recent soils. His detractors had argued that fossils were inorganic and could be found in any soils or rock. Steno knew otherwise – they were only found in layers of sedimentary rocks, and, comparing tongue-stones with the teeth in his dissection, he thought they were most definitely organic. It was an extraordinary scientific breakthrough, and all because of a white shark's teeth.

CHAPTER 3

SHARK BITE INCIDENTS DOWN THE CENTURIES

While the maritime folk of the Mediterranean region were familiar with the white shark and were in awe of it, if not sorely afraid of it, those plying the Atlantic coasts of north-west Europe rarely encountered such a creature, that is until they crossed the ocean and began to explore the shores of the Americas. Then, their worst nautical nightmares became realities: large sharks, including white sharks, were abundant and, as often as not, confrontations were far from pleasant experiences.

One of the early settlers was Bartolomé de las Casas (1484–1566), a Spanish conquistador turned Dominican friar who is known for the way he exposed the atrocities carried out by the invaders on the indigenous people. Not so well known is that he also wrote about sharks, including the description of a shark attack.

In 1502, his first port of call was the island of Hispaniola, now divided between the Dominican Republic and Haiti, followed in 1511 with the invasion of Cuba. His fellow conquistadors were keen on making a fast buck or two, gold and silver being the preferred treasures, but they were also keen to get their hands on pearls. The Atlantic pearl oyster *Pinctada imbricata* was relatively common around the islands, as it once was on coasts from South Carolina to Uruguay. However, to obtain them they forced their slaves to dive in unsafe waters. The pearl diver accompanied by a Spanish overseer would paddle their canoe away from the shore and head out into water about 6 to 7 metres (20 to 23 feet) deep. The diver was then ordered to dive to the seabed, where he would collect oysters and place them in nets draped around his neck until he ran out of air, when he would surface. He didn't hang about, as the most vulnerable moment is when the diver is trying to hoist himself into the canoe; that's when sharks strike.

Las Casas identified two types of shark that were dangerous to humans: 'tiburones', which probably included tiger sharks and bull sharks, both potentially dangerous species; and 'marrajos', with a large mouth that can swallow a man with one gulp, a dead ringer for a white shark, and it was the latter that the pearl divers feared the most. Las Casas recorded one event that illustrated why they were so frightened. He tells of a time when a diver and his overseer paddled out to deeper water. However, when the diver went down, he came face to face with a marrajo and returned to the surface super-quick. The overseer demanded to know why he had returned empty-handed, and the diver said he was in danger from a big fish, and he was afraid it would kill him. Undeterred, the overseer forced the diver back into the water and beat him with a stick for good measure. Sure enough, the marrajo was waiting and attacked the diver. He put up a bit of a fight, and there was a commotion in the water, but true to form the marrajo took him in one enormous gulp. The overseer, realising his precious diver had been swallowed by a shark, killed a dog and used it as bait to capture the shark. The shark took the bait, and the overseer was able to haul it back to shore and onto the beach. He called for people to come help him land the shark and they used axes and anything else that came to hand to kill it. Opening the shark's stomach, they found the unfortunate diver inside. He was still alive, but barely, and after a few gasps, apparently, he gave up the ghost. The report by las Casas was one of the first records of a shark attack on humans in the New World.

In these waters, a shark of such an immense size that it could swallow a person whole should not surprise shark historians; after all, one of the largest white sharks ever caught was landed right here, at Cojímar, a small fishing village on the north coast of Cuba, not far from where Ernest Hemingway penned his Nobel Prize-winning novella *The Old Man and the Sea*.

The year was 1945 and local fishermen reported that their fishing gear, set to catch marlin and swordfish, was being destroyed by something large and mean. They decided to do something about it, so on a calm June day, a few of them sailed out in their 4-metre (13-foot)-long skiff and joined other fishermen trying to catch the precious billfish. They were out in the ocean, on the edge of the Gulf Stream, but since dawn, it had been eerily quiet. Even though they had baits of fresh ballyhoo halfbeak, nobody was catching a thing, and at about nine o'clock they saw why. A huge dorsal fin was cutting through the water, the biggest they had ever seen, and it was heading for the skiff. They knew immediately that this was a white shark, and this was clearly why they were no large fish about. They had all vamoosed. Somehow, they had to catch the shark, which would not only remove the problem, but also give them a prize fish that they could sell at market. They tossed bait and chum into the water to keep the fish interested, and attached a half tuna, which had been sliced up by a smaller shark the previous day, to a shark hook. The hook was attached to a wire leader and hundreds of metres of silk rope. The bait was placed gently in the water and the shark reacted immediately, first swimming alongside the skiff and then grabbing the tuna. It was hooked. It shot off at speed, the line trailing behind. To slow it down, they had attached several 'palangres'. These are small wooden rafts used when catching swordfish at night. They enable the fishermen to have many lines in the water at the same time – a type of long-lining, but now the drag of the palangres would hopefully tire out the shark. However, it dived deep and towed the palangres around the ocean for several hours in a scene reminiscent of *Jaws* – the bit with the barrels. There was nothing the fishermen could do but to hang on. Then, it eventually broke the surface. Several of the crew hauled in the rope so they came closer to the shark, and others prepared a harpoon and prayed. This was the most dangerous part.

It took an hour to haul in the rope, and peering down into the clear blue water, they could see the shark at a depth of about 20 to 25 metres (65 to 80 feet) under the skiff. It was huge – hearts jumped immediately into mouths. What had they taken on? When the shark was within 6 metres (20 feet) of the boat, the excitement started. The shark headed straight for the skiff and its six fishermen and slammed into the keel, knocking the boat sideways. Then, it turned and took some more out of the keel and a chunk out of the rudder, so thousands of slivers of wood floated on the sea's surface. At the next charge, one of the fishermen readied the harpoon – a wooden pole with a bronze tip – and plunged it into the shark, just behind the head.

When they finally got the shark back to shore, the entire village – men, women and children – came to see 'El Monstruo de Cojímar'. They climbed all over it and then posed for photographs taken by a French reporter from *Le Monde*, who happened to be on holiday in Havana. It was quite an event, and the size of the shark turned out to be staggering. It was a female estimated to be 21 feet (6.4 metres) long and to weigh about 7,000 pounds (3,200 kilograms). The liver alone weighed about 1,500 pounds (680 kilograms), surely a record breaker at the time, but it was never registered.

Coincidentally, in 2019 and again in 2021, two white sharks, Caroline and Miss May, tagged by OCEARCH – a US-based non-profit organisation that generates critical scientific data on keystone marine species, including white sharks – were present in the same part of the sea on the north coast of Cuba at about the same time of year as the Cojímar shark, so the scientists involved were hoping that their two tagged sharks did not suffer the same fate (see also Chapter 6). Mercifully, they didn't.

Back in the time of las Casas, other seafarers began to arrive from Europe, and they were in for a bit of a shock. They had tended to fish rivers and inshore waters, which were filled with fish, rather than brave storms on the open sea like the Cuban fishermen. Aside from the basking shark, which is the second largest fish in the sea but a harmless filter-feeder, they rarely came across large sharks like the white and none of the tropical species, such as tiger sharks and bull sharks. However, when the privateers, like Sir John Hawkins, Sir Francis Drake and Sir Walter Raleigh, known collectively as the 'Sea Dogs', headed across the Atlantic to plunder Spanish ships in the Caribbean and other tropical and semi-tropical waters in the Americas, they very quickly became acquainted with the elasmobranch communities, especially those with species that have a tendency to hurt or kill people. It led sea captains like Hawkins to bring the scary sharks they had caught back to Britain, along with new names. Until then, they had been known generally as sea dogs or dogfish, but the crews adopted the names they had learned from the local Amerindian people. The Spanish sailors took the word *tiburón* from the Carib Indians, which the English also used for a while, but in the later sixteenth century, the English borrowed the word *xoc* from the Mayans of Mexico, which is thought to have led to the English word *shark*.

It was also about this time that the first shark attack report in the English language was published. The observation is from an anonymous officer, as most ratings could not write, and it tells how he was sailing from Portugal to India in 1580, when one of the crew fell overboard during a storm. Other members of the crew tied a rope to a wooden block and tossed it into the sea close to the unfortunate sailor. The man grabbed the block and those on board quickly hauled him towards the ship but, when he was 'half the range of a musket shot' away, a shark attacked. The officer wrote: 'There appeared from beneath the surface a big monster known as tiburon; it rushed at the man and cut him to pieces right before our eyes. It was certainly a terrible death.'

With ships sailing more exotic routes, their logs began to fill with many reports of shark attacks. Not all were so gruesome. The captain of the *Ayrshire*, for instance, fell overboard. A shark immediately appeared, just as the captain's dog jumped in the sea to save him. Both man and dog reached safety, but the dog's tail was cut in half.

At about the same time as las Casas was writing about the Indians of the Caribbean, the noted Spanish historian, botanist and soldier Gonzalo Fernández de Oviedo (1478–1557) was pulling together the first natural history of the New World. With the title *Sumario de la Natural Historia de las Indias*, the book was published in 1526 and in it he drew attention to the great biodiversity in the region, including many species of sharks. He also put forward an explanation as to why white sharks (marrajos) were so numerous in the coastal waters of the islands. He felt that it might have had something to do with one of their principal foods also being abundant. White sharks of a certain size – not youngsters and not the largest mature individuals, but the ones in between – have a predilection for pinnipeds, and the Caribbean was then awash with monk seals. In his fascinating, prize-winning essay 'Historical Knowledge of Sharks: Ancient Science, Earliest American Encounters, and American Science, Fisheries and Utilization', veteran shark researcher José Castro, of the National Oceanic and Atmospheric Administration (NOAA)'s National Marine Fisheries Service in Florida, wrote that white sharks frequented these waters with their monk seals just as they frequent the waters around seal and sea lion rookeries elsewhere in the world today. Sadly, the Caribbean monk seal is no longer with us. It was declared extinct by the USA in 2008, after an extensive five-year search for specimens. The white shark, though, still occurs in the region and tracking studies show that it is a frequent visitor, albeit not very close to the coast, having probably switched to other foods, such as swordfish, marlin, dolphins and maybe even dead whales that are in deeper waters.

Back in the sixteenth and seventeenth centuries, the pearl-fishing industry gathered pace and, in the eighteenth century, when the Spanish and other overseers ran out of local Amerindians to dive for them – they had succumbed mainly to diseases brought from Europe to which they had no immunity and had died out – they turned to African slaves instead. Antonio de Ulloa, the governor of Florida and Louisiana, toured the Spanish possessions and drew attention to the intensity of pearl fishing on many of the islands, such as those in the aptly named Archipiélago de las Perlas off the Pacific coast of what is now Panama, where the native population was wiped out within two years of the Europeans arriving. At the time of Ulloa's visit to the islands, boats were going out, each with at least eighteen to twenty slaves and an overseer. Each diver was tied to his boat by a rope and carried a rock to get to the seabed quickly. A foreman slave oversaw the operation and looked out for sharks. Should one appear, he would tug at the ropes to alert the divers and even jump in with a weapon to see it off. Nevertheless, divers lost arms, legs and even their lives.

While Ulloa's publication *Viaje a la América Meridional* saw the light of day in 1748, a couple of decades later a British scientist – Welsh actually – wrote a short essay on the white shark in his four-volume encyclopaedia *British Zoology*, published in 1776. The Welshman was Thomas Pennant (1726–98), and he was far from complimentary about the white shark.

It 'grows to a very great bulk', he wrote, 'to the weight of four thousand pounds [which suggests Pennant's white sharks were pretty big]; and … in the belly of one was found a human corpse entire, which is far from incredible, considering their vast greediness after human flesh.'

Well, after that opening statement, it was all downhill from there.

'They are the dread of the sailors in all hot climates,' he continued, 'where they constantly attend ships in expectation of what may drop overboard; a man that has that misfortune perishes without redemption: they have been seen to dart at him, like gudgeons [a bottom-dwelling freshwater fish] to a worm.'

And like his predecessors he couldn't resist playing the slave card, but with the most gruesome story.

A master of a Guinea ship informed me, that a rage of suicide prevailed among his new bought slaves, from a notion the un-happy creatures had that, after death, they should restore again to their families, friends and country. To convince them that they could not reanimate their bodies, he ordered one of the corpses to be tied by the heels to a rope, and lowered into the sea, and tho' it was drawn up again as fast as the united force of the crew could be exerted, yet in that short space the sharks had devoured every part but the feet, which were secured at the end of the cord.

Pennant went on with his white shark takedown. 'Swimmers often perish by them; sometimes they lose an arm or leg, and sometimes are bit asunder, serving but for two morsels for this ravenous animal.' But he also continued with a more scientific account:

> The mouth of this fish is furnished with a sixfold row of teeth, flat, triangular, exceedingly sharp at their edges and finely serrated … This dreadful apparatus, when the fish is in repose, lie quite flat in the mouth, but when he seizes his prey, he has power of erecting them, by the help of a set of muscles that join them to the jaws.

Here, Pennant has identified correctly the white shark's protrusible jaws, which are thrust forward at the moment of biting. However, he also repeated Pliny's false notion that, as the shark's mouth is on the underside of its head, it must have to turn on its back to bite, not realising the 'erect' teeth or protrusible jaws enable the shark to push its jaws forward, beyond the snout, and bite while still the right way up.

Despite this, many of Pennant's notes were fairly accurate. It's his sensational accounts of the white shark attacks that rankle. They gave the white shark a bad name and it's stayed that way ever since. The odd thing, though, is that the white shark description is in a book with the title *British Zoology*. Did Pennant know something that we don't? This was repeated again in Jonathan Couch's four-volume opus *A History of the Fishes of the British Isles*, published between 1860 and 1865. The Cornish naturalist wrote that, in the West Indies, the white shark 'is the dread of sailors who are in constant fear of becoming its prey when they bathe or fall into the sea'.

Most of these accounts were circulated among seafaring people as, during the eighteenth and nineteenth centuries, sharks didn't feature very strongly in the world's newspapers, and there were only a few scattered reports of shark attacks on humans. A probable reason was that many people could not swim at that time, so there were fewer in the water. There were, however, a few tragedies involving those who could.

One of the most famous was the plight of Sir Brook Watson, who, in 1749, was just fourteen years old and a crewman aboard a ship anchored in Havana harbour, so he was in a known white shark region. One day, he went swimming in the harbour and was attacked by a large shark, species unknown but which could have been a white shark, tiger shark or bull shark. The shark bit twice, taking the flesh away from his lower right leg before returning to bite off the young man's right foot. His crewmates had seen the attack and immediately went to the rescue. Watson spent three months in a Havana hospital but had to have his leg amputated. Later in life he went into politics in London and commissioned a picture of his ordeal by the American artist John Singleton Copley. Copley probably had not seen a shark in real life, so his depiction is not anatomically correct, but nevertheless it is recognisable as a large and ferocious shark. The picture had a huge impact on English society, probably one of the first emanations of a phenomenon that became known much later as the *Jaws* Effect (see also Chapter 13).

Sharks were not only having a go at people, but also attacking the vessels in which they were sitting. Two events, for example, that involved a white shark bumping and capsizing a boat occurred in 1730 in Boston harbour and 1771 somewhere on the New England coast. Alexander Sampson was the first victim. He was on a leisurely boating afternoon, when the shark attacked his boat and tipped him into the water. Then, it ate him. The second was a man in a boat with two others. Similarly, a shark ran into the boat, knocking the man into the water and devouring its victim.

The strange thing about both these incidents is that this behaviour was something the First Nations people on both the east and west coasts of North America (see also Chapter 1) recorded and were afraid of, but nowadays you infrequently hear of white sharks bumping boats and knocking their occupants into the water. They simply mouth anything metal, like an outboard motor, because their electromagnetic sense is overwhelmed, but they don't actually attack. Were the sharks more aggressive in those days? We'll never know, of course, but it's an intriguing observation, and even though it's not common nowadays, there have been a few more recent attacks on boats.

In 2014, a white shark attacked the craft of two kayakers off Plymouth, Massachusetts, knocking the occupants into the water. They were traumatised but otherwise unharmed. And in July 2021, at Moss Landing Jetty, Monterey, California, a kayaker was hit from below by a white shark and also ended up in the water. He was bumped by the shark and then his kayak was grabbed and pushed about 10 metres (33 feet) away. Fortunately, the kayaker survived without injury. Elsewhere in California, kayakers were tossed into the water at Shelter Cove, Humboldt County, in August 2020, at Pigeon Point, San Mateo County, in January 2017, at Santa Cruz in November 2017, and at San Carlos Beach, Monterey, on 18 March 2017, so sharks bumping boats is alive and well. All the people survived.

In the early nineteenth century, Bristol, Rhode Island,

became the centre of attention when a young boy swam from a boat towards the shore but never made it. His body was found later, minus all of its limbs. A white shark was thought to have been the predator responsible. The year was 1816, when shark bite incidents on the north-east coast of the USA were mercifully few.

By the mid-nineteenth century, however, attention switched to the antipodes. In January 1852, nineteen-year-old Johnny Balmer, a young musician with a military band, was bathing with a bunch of his bandmates. It was a warm summer's day in Wellington, New Zealand, and the band had been entertaining the citizens celebrating their city's anniversary. When the rest of the bathers decided they had had enough, they swam back to the shore, leaving Johnny to linger a little longer. He was about 60 metres (66 yards) out, when another young man passed close by in a boat. He asked the soldier if he wanted a ride back.

'No thank you,' he replied, 'I think I could keep swimming for an hour yet, so I'll go back the way I came.'

'As you will,' said the sailor, and the boat carried on.

At that moment, the boatman saw the large dorsal fin of a shark – probably a white shark – appear not 30 metres (33 yards) away. 'A shark, a shark!' he cried. 'Johnny, come to the boat!' But it was too late. The shark slammed into the soldier, and he screamed. It bit on his thigh and calf and dragged him below. The sea turned crimson. The captain on a nearby ship estimated the shark to be about 4–5 metres (13–16 feet) long. Johnny broke free and made it to the boat, but he had lost an enormous amount of blood and bled to death before they could get him to shore.

An inquest concluded that 'the deceased had been bitten by a shark whilst bathing', a remarkable thing at that time, because the blame was pinned on all manner of beasts and, like the mayor in *Jaws*, nobody wanted to accept that there could be dangerous sharks in their waters to drive away holidaymakers. It was something of a breakthrough, and the only fatal shark bite incident in Wellington's history.

Shark attacks in New Zealand waters are thankfully rare, but they do occur. South Island is the realm of the white shark, so you'd expect more fatalities than around North Island, where there are more incidents but generally non-fatal. Around South Island, from 1888 to 2014, for example, there were twenty-five shark bite incidents of which six were fatal, probably the result of white shark attacks, hotspots being St Clair and Moeraki in the south-east where nine attacks occurred.

One incident in April 1992, however, was totally unexpected. The location was Campbell Island, a sub-Antarctic island about 660 kilometres (410 miles) south of South Island. A meteorologist stationed on the island was snorkelling with friends when he was hit by a white shark. It ripped off his arm, but the man survived thanks to the courageous rescue work of one of his colleagues and a local helicopter pilot, both of whom received bravery medals. That white sharks reach so far south, right on the edge of the icy and windy Southern Ocean, came as a surprise to the meteorologists, unaware of the special heat exchange system (see Chapter 6) that enables the white shark to retain heat and which means it can survive in cold subpolar waters where most other sharks choose not to go. New Zealand sea lions and southern elephant seals are a likely attraction.

To return to our timeline, the early twentieth century was dominated by events at Matawan Creek in New Jersey (see Chapter 4 for the full story), and it was these shark bite incidents, above all others, that sealed the white shark's reputation and raised the notion of a 'rogue shark' that, having had a taste of human flesh, thereafter sought it out, as featured in the *Jaws* book and film. However, by all accounts, white sharks are unlikely to seek us out because they like energy-rich blubber from marine mammals, especially seals and whales, and even the most obese of us can't compete with them. Nevertheless, as soon as there is a cluster of shark bite incidents in a particular area, the press immediately seizes on the idea of a rogue shark at work.

The idea of a shark seeking out human prey has been around since the 1930s, but it came to prominence in 1950, when the Australian surgeon Sir Victor Coppleson (1893–1965) published in the *Medical Journal of Australia* his paper on shark attacks. Coppleson had been in Europe for the two world wars, but when he returned to Australia there were many more recreational swimmers and surfers in Australian waters, and consequently many more shark bite incidents recorded, than there had been before he left. He collected together as many records as he could find and attempted to analyse the behaviour of these fish. He put forward the notion that sharks that chomp on humans are lone 'rogues' and that they have developed a taste for human flesh. They stake out stretches of coast, a river, a harbour and other places where people are likely to be and wait for a chance to strike. As evidence, he quoted several historical shark bite incidents in Australia and elsewhere. In *The Narrative of the Expedition to Botany Bay,* published in 1793, for example, Lieutenant General Watkin Tench, a British marines officer, described how a woman of the First Nations was bitten in two off the coast of New South Wales. This was in about 1790 and certainly sounds like the horrific work of a white shark, a species which was and still is present in the area. It was the first recorded shark bite incident from Australian waters.

Coppleson also put forward multiple events that he said possessed all the hallmarks of a rogue shark. In 1922, a series of attacks occurred at Coogee, near Sydney. 'The first attack,' he wrote, 'was followed by another a month later and two more within the next three years.' All four were within about 2 kilometres (1.2 miles) of the coast. Then, an attack took place in 1928 at Bondi, also near Sydney, with another in less than a year, a third a month later and, after ten days, a fourth about 5 kilometres (3 miles) away. The first three occurred at the same beach. And, between 1934 and 1936, 'there were five attacks on the northern Sydney beaches', while there had been none in previous or following years.

As further evidence, Coppleson mentions that South African beaches had spates of attacks. In 1944, before meshing was introduced, the beaches in Durban had five incidents within eighty-one days, and in 1946, there were four attacks in twenty-three days. Between Margate and Karridene, there were five attacks in twenty-three days, and Winklespruit had two attacks in three days in 1962. He attributed the attacks to rogue sharks. 'Each series', he wrote, 'is the work of a single shark – a rogue shark – which maintains, even for years, a beat along a limited stretch of shore.'

The sharks involved in these incidents are 'abnormal',

according to Coppleson, the so-called 'man-eaters' that had been described so vividly in the scientific literature during the late nineteenth century. Other sharks are unlikely to bite people, he said, and so should be considered 'normal'. His conclusion was that a rogue should be 'hunted until it is destroyed'. Bad news for sharks. In 1958, Coppleson published a more popular account in his book *Shark Attack*, and the phrase 'rogue shark' became fixed in the English language. Since then, many of his ideas have been rejected by today's scientists. Human-flesh-loving sharks may be the stuff of nightmares, but the reality is that they are from the realms of science fiction rather than science fact.

It's quite possible, however, although exceedingly rare, for a white shark to bite several people in a particular area over a short period of time, in a kind of cluster pattern, but this is not necessarily a rogue with a desire for human flesh. It's just a shark taking advantage of what's available rather than seeking out humans especially. This is probably what occurred at Matawan Creek; the culprit, according to George Burgess, former director of the International Shark Attack File, being a white shark (see also Chapter 4).

Post-Matawan, holidaymakers all along the east coast of the USA were understandably nervous. However, things quietened down for the next couple of decades, with few incidents, that is until 25 July 1936, when newspapers were filled once more with news of a deadly shark attack.

The victim was sixteen-year-old Joseph Troy Jr, a summer visitor from Dorchester, a neighbourhood of Boston. At between three and four o'clock on a Saturday afternoon, with the weather warm and sunny, he and a family friend, Walter Stiles from Boston, were swimming out to a catamaran that had just picked up its moorings off a beach near Mattapoisett, Massachusetts. The two swimmers were about 10 feet (3 metres) apart, Stiles adopting a leisurely sidestroke and Troy swimming with a crawl and causing a considerable commotion in the water. When about 165 yards (150 metres) from the shore, where the water was 10–15 feet (3–4.5 metres) deep, Troy was attacked on his left side, the opposite side to which Stiles was swimming, by a relatively small but incredibly vicious white shark, 6 feet (1.8 metres) long. It dragged him below by the left leg, but the owner of the yacht was able to manoeuvre his dinghy close enough to catch Troy, and with the help of Stiles, pulled him aboard. On the shore, several neighbours improvised an old door as a stretcher, and he was driven to St Luke's Hospital in New Bedford. Entering the operating theatre, the young man said to the surgeon, 'It was not the bite that worried me, but the thought of being dragged down by the shark.' Those were his last words. While the surgeon was amputating his leg, he died in the operating theatre.

When the word got out, it was Matawan Creek all over again. The *Sunday Standard-Times*, for instance, ran the front-page headline 'Shark kills swimmer in Mattapoisett Cove'. However, it was the only fatal shark attack on this stretch of coast for more than eighty years. In the summer of 2018, all that changed. Engineering student Arthur Medici was boogie boarding at Newcomb Hollow Beach, one of the beaches in the desirable resort of Wellfleet, when he was hit by a white shark. It was a sunny midday in September, but a cloud descended immediately over Wellfleet, site also of an attack on a paddleboard the year before.

Local people rushed to help the victim. One woman, caring not for her own safety, dragged the injured Medici from the surf and away from the shark. 'I saw the blood, I went in,' she told reporters. 'A wave came up over Arthur's face. I got underneath and pulled him up on top of me.' But it was to no avail. It's thought he had already bled to death before they could get him to the shore.

Shark researcher Greg Skomal, of the Division of Marine Fisheries, told the *Cape Cod Times* that it was probably a case of mistaken identity. The shark was really after seals. Seals tend to hang about in the shallows, believing wrongly that the sharks won't be able to get at them. The sharks, however, cruise the slightly deeper water channels between sandbanks and, instead of approaching their target from below and behind, as they do elsewhere in the world, they rush the victim laterally. It's behaviour that seems to be unique to Cape Cod. Also unique is that the seals, sharks and people are on the same beaches, so there's bound to be conflict.

Indeed, it was the second attack that summer, the first being a non-fatal incident at Truro, about 6 kilometres (4 miles) north of Newcomb Hollow Beach, in August. A 61-year-old man was attacked but, when the shark clamped down on his leg, he fought off his assailant by punching it in the gills. He managed to swim back to shore and then yelled for help. Off-duty nurses and other medical professionals helped to stem the bleeding and the man was carried up to the car park. He was airlifted to Tufts Medical Center in Boston, and after being put in a coma for a couple of days, undergoing six surgeries and being given 12 pints (5.7 litres) of blood, he lived. 'The pain was really excruciating,' he told Associated Press. 'I remember the helicopter landing and then nothing for two days.'

At this time, frequent shark sightings around Cape Cod resulted in an unparalleled number of beach closures. The National Parks Service, which administers many of the picturesque beaches in the area, indicated that, in summer 2018, beach closures were double the annual average. The sharks were back.

One reason for the hiatus was that, during the previous few decades, the sharks were just not coming to New England. Seals and people have happily coexisted in New England since at least 2500 BCE, but when seal bounty hunters began plying their nefarious trade around 1880, they decimated its seal populations, so there was nothing to make a visit worthwhile. The sharks were out there on migration but just passed by on their way north to Newfoundland and Labrador where there were still a few seals. When conservation laws made sealing illegal in the USA – 1965 in Massachusetts – the seal population slowly built until it came to the point when the sharks had something to eat once more. It went from local extirpation of grey seals on the north-east coast of the USA and a few hundred harbour seals in the Gulf of Maine during the early 1990s to an estimated 30,000 to 50,000 grey seals in the Cape Cod area alone today. Shark numbers are also impressive: 320 white sharks were spotted off Outer Cape and the inner part of Cape Cod Bay between 2014 and 2019. It's not known, of course, how many of these were counted more than once, but 151 were tagged by shark scientists, so there were at least that number in the area. A similar pattern has been observed in South Australia, where fur seal numbers have been increasing.

All this activity has naturally increased the amount of shark research, but it hasn't always been this way. Despite evidence of sharks biting people going back for thousands of years, during the 1930s, the question 'Do sharks attack people?' was still being asked. The eminent ichthyologist Eugene Willis Gudger (1866–1956), of the American Museum of Natural History, who wrote more than 300 articles on fish, asked that very question. He acknowledged, somewhat reluctantly it seems, that sharks do, indeed, bite people, although he attributed most attacks to the barracuda, a bony fish possessing a mouth filled with needle-sharp, fang-like teeth which could do damage should it grab an arm. There are few fatalities. In 1947, a diver off Key West died, thought to be the result of a barracuda attack, and in 1957 another fatality occurred off the coast of North Carolina. More usually any victim is left with serious lacerations, like the free diver who encountered a barracuda off Pompano Beach, Florida, in 1960, and came away needing thirty-one stitches.

The first half of the twentieth century was dominated by the two world wars. While the war at sea during World War I was predominately in the North Atlantic, where most casualties at sea succumbed to the cold water at higher latitudes, in World War II, downed airmen and the crews of sunken ships would survive for days or even weeks while waiting for rescue, floating about in the warmer waters of tropical, subtropical and warm temperate seas. It was this warm water group that became the target of sharks, and the authorities, especially in the USA, became very aware of the dangers to military personnel. The result was a period of intense research to find a solution, although the initial stimulus was more by accident than design.

The key player was the anthropologist Henry Field (1902–86), then at the Field Museum in Chicago. In 1941, he became a special advisor to President Franklin D. Roosevelt on post-war migration and, the following year, found himself in Trinidad. Here, he heard hair-raising stories from the crews of merchantmen who had been torpedoed and harassed by sharks. It affected him profoundly, and he lay awake at night trying to work out what to do to change things. His solution? To create a chemical to repel sharks. Sounds simple, but it was far from it. For one thing, finding a substance that would repel sharks effectively was a huge undertaking, and for another, the US Navy seemed reluctant to admit that sharks were a problem. Field mentioned that one naval officer claimed there was no record of US naval personnel having been attacked by sharks. Well, that attitude changed dramatically on the night of 23 July 1945 when the USS *Indianapolis* was torpedoed and sank. Of the 800 who went into the water, just 316 survived. Sharks got many of the rest. Quint – the fictional shark fisherman in *Jaws* – was there. However, it's unlikely that white sharks were among the attackers; probably oceanic whitetip and blue sharks were the perpetrators, species known to be ultra-inquisitive and which usually turn up at shipwrecks.

With everyone having woken up to the fact that sharks do attack people, especially downed airmen and the crews of torpedoed ships, the search for a shark repellent came on apace. The result was Shark Chaser, a substance based on the fact that sharks are repelled by other dead and rotting sharks. After several years of research, the scientists isolated the active ingredient – copper acetate – and put it in the repellent. It became standard issue for US military personnel, but there was one problem – it didn't work! Even so, it was issued up until 1976, when it was finally withdrawn. More recently, scientists have been lobbing cans leaking rotting shark flesh into the water to see how sharks react; alas, there is not a deterrent yet in sight.

Another interesting development was the Moses sole, which we met in Chapter 2, and which lives in the western Indian Ocean. Its ichthyotoxin, which serves to deter predators such as sharks, seemed as if it might be the basis of a personal shark deterrent, but nothing came of it. Now, research is focusing on the toxin's antimicrobial and other potential medical properties.

Since then, numerous shark repellents of one sort or another have been brought out, but none have proved to work satisfactorily. The thing is that sharks have evolved to be the way they are over millions of years. They're not going to be fooled by some upstart system that took a few weeks or months to develop. The most used deterrents today, especially by surfers, are ESDs, which stands for electrostatic discharge. These are designed to emit pulses of electricity that interfere with or overwhelm the shark's electromagnetic sensors in its snout and, in effect, confuse it to the extent that it will stay some distance away; at least, that's the theory. Some ESDs are better than others, and one device comes to the top of the pile as far as white sharks are concerned. In tests of five popular modern deterrents by Save Our Seas Foundation project leader Charlie Huveneers and his team, Ocean Guardian Freedom + Surf, an Australian product, was the only device to shown to have any deterrent effect on white sharks. A surfboard without the device, for example, had sharks coming to within 1.5 metres (5 feet), whereas with the device on board and switched on they didn't come closer than 2.6 metres (8.5 feet). However, in tests using a static bait, the ESD only prevented sharks taking the bait about 60 per cent of the time, which left 40 per cent when they did take it. The efficacy of the device appears to depend on how hungry the shark is!

An interesting development in South Africa with its white sharks is the fitting of lights on the underside of surfboards and kayaks. Like the bioluminescence organs of deep-sea fish, the lights blend in with background sunlight, effectively rendering the surf board invisible – counterillumination. As the white shark is an ambush predator that relies on seeing the silhouette of its prey at or near the surface, its ability to see the board is disrupted. In tests with cutouts of artificial prey, the lights reduced the rate of shark attacks, and the brighter the illumination, the more effective the deterrent. The only problem is it requires considerable battery power, and therefore extra weight, to work, so researchers at Macquarie University, Sydney, are experimenting with different light arrays to create a model that can also function correctly as a surfboard.

Aside from personal deterrents, sharks and people are sometimes kept apart by protecting a popular beach with large mesh shark nets or by drumlines with baited hooks. Both are effective but are not flawless. The sharks can swim around them. This happened in 2020 at Greenland Beach, Queensland, which was equipped with both shark nets and drumlines, when a shark fatally attacked a surfer. Sharks are also not the only ones that get caught. Sea turtles and humpback whales, especially, end up in shark nets, and if unable to get out, they drown.

Nowadays, nets are being replaced by camera-equipped drones. The drones fly along a route close to the surf break at a height of about 60 metres (195 feet) above the sea and parallel to the shore. When a shark is spotted, they drop down to about 30 metres and try to identify the species and its size. If it's perceived to be potentially dangerous, inshore craft head out and follow it, informing people on the shore of its whereabouts.

Drones have their limitations, of course. It's difficult to see below the waves if it's too windy or raining, for example, but generally they afford significant protection, particularly at busy beaches.

Whatever the way of avoiding being bitten, there's no getting away from the fact that nowadays, more swimmers and surfers are in the water than there ever were, and that drone technology has enabled us to see how close we are to sharks and how often. It's surprisingly common, yet the number of shark bite incidents recorded are infinitesimally small. Save Our Seas quotes less than one incident per million people every year and about six people per year die as a result of a shark attack. If you consider the many millions of people swimming or surfing all over the world, that's a remarkably small figure. Sharks, generally, are not really interested in us.

Nevertheless, they do bite people and, according to the International Shark Attack File, the white shark is top of the shark bite league table. By 2023, for instance, there were a total of 351 unprovoked bite incidents by white sharks recorded since the beginning of the twentieth century, of which fifty-nine were fatal, almost three times the number of its nearest rival, the tiger shark. Indeed, five of the ten fatalities from unprovoked attacks registered in 2023 were due to white sharks, of which there were three reported in Australia, one in California, and another in Mexico.

Today, there are four main locations globally where white sharks and people swim cheek by jowl: California and the Pacific coast of North America, the Cape Cod region of North America, the southern half of Australia, and South Africa, but there are some surprises. The presence of white sharks is not necessarily the makings of a horror story.

Southern California was the location of a two-year field experiment focused on the behaviour of juvenile white sharks. Using drones, the researchers from California State University Long Beach Shark Lab visited twenty-six Californian beaches and plotted where the young sharks were present. It turned out they were around people for 97 per cent of the time the experiment ran. Both humans and juvenile white sharks occupied the same section of sea, within about 500 metres (550 yards) of the shoreline where the water is less than 8 metres (26 feet) deep, and the sharks did not attack their neighbours. They seemed to live in harmony with them.

The two protagonists – people and sharks – did not overlap everywhere along the southern California shore. Instead, mixed-sized aggregations of more than forty individual sharks, with lengths between 1.5 and 3 metres (5–10 feet), remained together for weeks or even months, with a clumped distribution. Overlap was noticeable especially at two locations – Carpinteria Beach and Del Mar Beach, and here sharks came closest to stand-up paddleboarders more than other recreational beach users, such as surfers, bodyboarders, swimmers and waders.

The reasons for juvenile white sharks being close to shore are probably three-fold: firstly, it is safer to be inshore away from older sharks that could turn cannibalistic, which also removes any competition with them for food and living space; secondly, there's a plentiful supply of fish, the primary food for juveniles; and thirdly, they prefer the warmer waters close to shore, which is why the people are there too. It is thought that they are less likely to attack the people with which they share these waters because young of the year (the newly born sharks) and juveniles feed mostly close to the seabed on bottom-dwelling fish. Humans are considerably bigger, and surf-boards are the same length or longer than the sharks, so they ignore any surface-orientated activity. They are intimidated by objects of a similar size to themselves, so merely having a surfboard reduces the risk of being bitten by these juveniles.

However, we shouldn't be too complacent. The chances of being bitten by a juvenile white shark cannot be said to be zero, just very low; in fact, there was a bite incident in the spring of 2020 in which the victim saw a juvenile white shark hastily leaving the area. Whether it was the perpetrator of the attack, which resulted in minor lacerations, is unclear. What is clear, however, is that their older relatives are a different kettle of sharks.

Even so, unprovoked shark bite incidents here are rare: just 201 in California as a whole from 1950 until 2021, of which 178 are known to have involved white sharks. The figure since the turn of the century for southern California is just twenty incidents, which given that southern California has about 129 million beach-related visits per year, and climbing, is a pretty low number. Any attack, however, is sure to hit the headlines, with surfers most frequently on the receiving end. As white sharks grow, subadults being 3 to 4.5 metres (10 to 15 feet) long, they switch diet from fish to marine mammals, so are surfers the unwitting stand-ins for the subadult white shark's preferred prey, seals and sea lions? The sharks do seem to be confused by people. Attacking us, it seems, could sometimes be cases of mistaken identity.

To a white shark, say about 60 metres (200 feet) down and scanning above while in hunting mode, a person swimming or surfing and silhouetted against the light at the surface resembles an aberrant seal. We're just about the right size, compared to a large seal, and anything behaving in an odd way is going to attract attention. The shark goes into attack mode, taking a bite where it thinks it can do the most damage. This is often towards the middle or rear of the board, so the shark seems to be hedging its bets. It then stands off waiting for the 'prey' to die or it moves away and hunts elsewhere because the food, especially the board, is not up to scratch.

It is this bite-and-spit behaviour that is partly responsible for fewer human fatalities. The statistics are surprising: many more people survive a white shark attack than are killed. The International Shark Attack File reveals that fewer than 17 per cent of unprovoked attacks have been fatal since records first began in 1580. The sharks appear to take an exploratory bite and then spit us out. Even the most obese of us doesn't have enough calories to make it worthwhile to press home the attack. As long as the flow of blood from ripped arteries can be controlled, the victim's chances of survival are remarkably good, which is surprising if you consider that this 1.5-tonne marine creature, with jaws filled with razor-sharp, serrated teeth is pitted against a 100-kilogram (16-stone) land-based primate flailing about in a medium that is alien. White sharks should be gobbling us up all over the place if we were a chosen target, but we aren't. White sharks are really not that interested in people; in fact, there's some evidence to suggest that even in fatal attacks, the sharks do not always consume the body.

The first 'attack' might not even be an attack at all but an exploratory bite to ascertain whether you're edible or not. White sharks bite all manner of non-edible items – fishermen's floats, kelp and other floating debris. Their teeth and gums probably act as mechanosensory receptors, similar to our sense of touch. The shark doesn't have fingers like we do, so the only way it can learn more about what it is taking an interest in is with its mouth; in other words, the bite might not necessarily be simply feeding behaviour. It's just curious. One incident from the west coast of the USA suggests that this could be a rational explanation for some so-called 'attacks'.

Jon Holcomb was a commercial abalone diver and in September 1974 he was diving solo at the Farallon Islands, although he was not completely alone. He was using a regulator valve linked via a pipe or downline to the compressor of a boat on the surface operated by his support crew. There were also numerous seals and sea lions in the water just 300 metres (330 yards) away. Holcomb was on the rocky seabed at a depth of about 10 metres (33 feet) when he noticed that the wealth of fish life that he normally saw wasn't there. He'd been down about an hour and a half and was swimming rapidly across the bottom in search of abalone when a white shark, estimated to be 4–5 metres (13–16 feet) long, first slammed into his side and then grabbed his right arm and shook him for about five seconds before releasing him. The shark then bumped him in the chest with its snout four or five times. Holcomb dropped his abalone iron, used to prise the shellfish off the rocks, and the shark grabbed his left arm. Again, it shook him for another five seconds, and *again*, let him go. As the shark swam away, Holcomb picked up his metal bar and struck it on the side. The shark swam no more than 5 metres (16 feet) before it turned abruptly and hit the diver on the chest again with its snout. As the shark made to go off, Holcomb grabbed the corner of its mouth to prevent it from biting him, and was pulled so fast through the water that his face mask slipped down to his chin. Looking up, he saw the light above, let go and headed for the surface. He popped up about 15 metres (50 feet) from his tender, yelled out and his colleagues came quickly and hauled him aboard. After a call to the coastguard, he was airlifted to a San Francisco medical centre, where the doctors attended to lacerations on both arms and one thigh – all relatively minor when you consider that Holcomb had a close encounter with a hefty white shark equipped with a mouthful of seriously sharp teeth. Although, at the time, it must have seemed a violent affair to Holcomb, the shark was remarkably gentle, not at all like the slam-bam attacks on pinnipeds. It was as if the shark could see that the diver was not a seal or a fish, but nevertheless decided to find out what it was. It undertook what one commentator described as a series of 'tactile experiments' to determine what exactly it had come across.

Another thing that struck scientists about the Holcomb incident is that it occurred on the seabed and not at the surface, where you would expect a white shark bite incident to take place. As a sneaky predator, the white shark more usually positions itself below and behind its target, ready to rush at it and take it by surprise. Holcomb was certainly taken by surprise but was not hit with a killer bite. The conclusion by many scientists is that the incident appears to have been exploratory rather than predatory.

Sharks have good daytime eyesight so, depending on the clarity of the water, they can pick up visual cues from some distance away. Also working in the Farallon Islands, researcher Scot Anderson revealed that white sharks will examine anything floating on the sea's surface. First, they cruise around it while carrying out a visual inspection, and then, if nothing untoward happens, they mouth and nibble the object as they try to establish what it is; again, more evidence that white 'attacks' on people could, in some cases, be simple curiosity.

That curiosity, though, can be either less damaging or devastating for the target, which is illustrated by the next three cases. In one, near the New South Wales coastal town of Eden, a diver on the seabed was grabbed by the head and dragged for several metres before the shark let go, while the other two – one at Fort Bragg, California, and the other at Mexico's Tobari Bay in the Gulf of California – were in similar scenarios, collecting shellfish from the seabed, except the victims were decapitated. It just goes to show that even an investigative bite from a large white shark might not be a gentle affair and could well prove to be severely debilitating or even fatal.

There is also the shark's character to take into account. White sharks are ostensibly intelligent fish, but just like humans, some are probably cleverer or dimmer than others. The more intelligent individuals might like a challenge and do their investigatory bites, while the less bright sharks might be totally confused and attack willy-nilly. It was probably one of the smarter sharks that turned up at Spanish Bay, California, in December 1981.

Lewis Boren was surfing here but disappeared. The following day, his surfboard washed ashore having had a large bite taken out of it, probably the work of a white shark. The day after that Boren's body washed up with a similar single shark bite. The rest of his body was intact. There were no signs that the shark had fed on it, and the perpetrator was a big one. From the bite marks on the board and body it was estimated to be in the region of 7 metres (23 feet) long. It could easily have swallowed Boren whole, but it didn't. It simply discarded him after the first exploratory bite. If other surfers had been nearby, he might have survived, but the fact that he died indicates, perhaps, that he was alone and/or in a remote place.

In fact, the location of an attack might well be a significant factor in whether someone lives or dies, as a spate of white shark bite incidents Down Under during 2023 and 2024 appear to illustrate. Three of four deaths in Australian waters in 2023, for example, were from white sharks patrolling the Eyre Peninsula, South Australia, where seals were the intended targets, but surfers were caught in the crossfire. In May 2023, a schoolteacher was bitten fatally by a white shark at Walker's Rock Beach near Elliston. His body was never recovered. This was followed in November the same year by a 55-year-old surfer who died at Streaky Bay. A second attack at Elliston, on a 64-year-old surfer, also a schoolteacher, occurred in January 2024. The man was grabbed by the leg but was treated successfully at local and Adelaide hospitals. It prompted the local authorities to consider a shark cull.

Off the remote Ethel Beach on the Yorke Peninsula, to the west of Adelaide, a teenager was surfing alongside his father when he was attacked by a white shark, but, alas, he did not survive. It was one of eleven fatal attacks in South Australian waters since 2000, and the one thing that is common to all these incidents, aside from being encounters with white sharks, is that all the locations are remote. Statistically, it seems, you are more likely to be bitten off a remote beach than at one of the more popular resorts. Also, at a beach without the usual safety features, like lifeguards and paramedics, the shark bite incident is more likely to be serious, even fatal, the danger being the victim bleeds to death before help arrives.

The benefits of surfing where there are people and therefore potential rescuers was illustrated vividly in 2009 by an horrific encounter involving thirteen-year-old Hannah Mighall and a white shark. Mighall was surfing alongside her cousin, 33-year-old Syb Mundy, at Binalong Bay in the south of Tasmania's Bay of Fires, where the water is spectacularly clear. The two were waiting for a wave when something grabbed the girl's leg. It felt gentle at first, but soon there was a force that pulled her off her board and into the water. What followed was far from gentle.

The shark lifted her and then pulled her below the surface. When she popped back up her leg was still in its mouth. Mundy came alongside and was hammering the shark as hard as he could on the side of the head, but it didn't seem to have any effect; but then the shark let go. Instead, it grabbed Mighall's board and pulled it below. Unfortunately, Mighall was still attached by the surfboard leash and so she was pulled below a second time. The shark, meanwhile, had bitten right through the fibreglass and foam, taking out a huge chunk. Then, somehow, Mighall was free. Her cousin grabbed her and pulled her onto his back, and he paddled as fast as he could towards the shore. Ironically, earlier that morning, Mighall had been practising surf rescues with her Surf Life Saving club. She had been the victim, but now she was being rescued for real.

Just then, a wave came. Mundy shouted: 'We've got to catch this one as our lives depend on it.' The shark circled the two surfers continuously and then followed them using the same wave right into the shore via a deep-water runnel, its dorsal fin clearly visible above the surface.

The beach had not been busy that day, but fortunately there was a doctor and a nurse who rushed to Mighall's aid and provided life-saving first aid until an ambulance arrived. Had they been surfing on a more remote beach, where there was nobody, the outcome could have been very different. However, Mighall lived to surf another day, although the curved scar shaped by the shark's mouth is still with her. That scar and the bite marks on the surfboard revealed that her shark was probably a female white shark about 5 metres (16 feet) long… in other words, a seriously big shark.

CHAPTER 4

JERSEY SHORE, 1916: BIRTH OF *JAWS* AND THE ROGUE SHARK

In the summer of 1916, New York City was in the throes of a polio epidemic, known then as 'infantile paralysis'. It was a little-understood disease that could severely handicap a child or cause their death, and, by the end of June that year, polio had already killed 165 children in Manhattan alone, and there appeared to be no respite. At the same time, the Great War was raging in Europe and staggering numbers of casualties in the trenches were recorded day after inglorious day in this hard-to-understand war across the Atlantic; yet a series of events along the New Jersey coast pushed all those horrors off the front page and came to dominate newspaper headlines in the USA for several weeks.

It all started on Saturday 1 July at Beach Haven, New Jersey, a nineteenth-century upmarket resort facing the Atlantic Ocean on Long Beach Island. It was known locally as 'Queen City' because it was the summer home to the middle classes who lived in Queens Village, a suburb of Philadelphia. The 24-year-old Charles Epting Van Sant, son of a businessman and one of those middle-class summer exiles, had just arrived on a very stuffy and crowded train. He, and most of the other passengers, had experienced an especially unpleasant journey. The summer of 1916 was unusually hot and humid. Day after stifling day, the temperature soared, and, at a time when powered air-conditioning units were not widely available, people were simply dropping like flies from heat exhaustion. Factories and businesses were shut down because of the oppressive heat. So, impatient to cool off in the sea, Van Sant left the rest of his family unpacking, donned swimming trunks and robe and headed for the ocean.

Van Sant was a strong swimmer. In no time at all, he was 100 metres (110 yards) out and turning for a leisurely swim back to the shore. On the beach, crowds of holidaymakers were searching for the best places to put their fold-up chairs and windbreaks, when they were suddenly stopped in their tracks. Women screamed. Parents quickly ushered their children from the shallows. They had all seen the black fin of a large shark heading straight for the young man. A man yelled to Van Sant to get out of the water, but he heard and saw nothing. Little more than 15 metres from the shore, the gap between the two narrowed, and the beach fell eerily silent; then, the water churned red.

United States Olympic swimmer Alexander Ott was on the beach. He immediately dived into the water and swam towards the rapidly spreading stain. The dark shadow beneath the surface turned towards him, but then thankfully veered away. Ott dragged Van Sant to the beach, the shark following. When he reached the sand, he discovered that the young man's thigh had been shredded and stripped of flesh. He died later that evening from shock and loss of blood.

The incident was quick to reach the newspapers. The *New*

York Times reported that Van Sant had been 'badly bitten in the surf here on Saturday afternoon by a fish, presumably a shark'. Surprisingly, no alarm was raised, even though the article also mentioned that 'large sharks' had been seen recently a few miles out. In their defence, the local authorities pointed out that there had not been a shark attack anywhere along the New Jersey shore in living memory... except there might have been.

Three years previously, on Tuesday 26 August 1913, a fisherman off Spring Lake, about 80 kilometres (50 miles) to the north of Beach Haven, caught a shark whose stomach contained a woman's foot complete with a tan shoe and knitted stocking. It was said that the shark must have scavenged the woman's body after some accident, for at the time it was genuinely thought that sharks would not attack a living swimmer; and who would go swimming in their shoes and stockings anyway?

Nevertheless, officials played down the Van Sant attack, saying that the shark had gone for a dog which was swimming close by, and had inadvertently bitten the young man. Assured that the experts had spoken true, the 4 July vacationers descended on the Jersey Shore as they always did, and there was no further news of sharks and no more shark attacks. Aside from a few nervous swimmers at Beach Haven, it was business as usual.

Spring Lake – site of the gruesome find in 1913 – was another resort for the well-to-do, especially for the barons of industry and members of New York and Philadelphia high society, many of whom lived in 'cottages' that were actually grand seaside mansions. Many Irish Americans settled in the area, and there was a fair sprinkling of Washington politicians with homes there too. However, on Thursday 6 July, talk was not of the Allied offensive, nor President Wilson's re-election prospects, nor the death of a young man at a less fashionable resort down the coast. Infantile paralysis was once more on everyone's lips, and the danger of it spreading to Spring Lake; and so on this day, few people were thinking about sharks.

The beaches were crowded, and just after lunch, 28-year-old Charles Bruder, bellhop at the fashionable Essex and Sussex Hotel, went for his afternoon dip. He was popular with both residents and visitors alike, and from the tips he earned he sent money to his ageing mother in Switzerland. By the time he reached the water, the tide had ebbed, so most people were sunbathing on the beach. Nevertheless, he waded in and was quickly beyond the lifelines, ropes and floats strung out in the water beyond which bathers were discouraged from going. These were patrolled by lifeguards George White and Chris Anderson. They would have shouted for other bathers to come back, lest they be washed out to sea, but they knew Bruder to be a strong swimmer and let him be.

Just then, a woman screamed. 'He has upset,' she cried. 'The man in the red canoe is upset!' White and Anderson immediately launched their rescue boat. There was no canoe, and the red colour that was spreading across the sea's surface was unmistakably blood. Bruder's head appeared, and he shrieked loudly, raising a bloody arm. White held out an oar and Bruder grabbed it. The two lifeguards hauled him in. His face was ashen white, and he was surprisingly light. As he was hauled from the water the rescuers could see why.

'Shark got me – bit my legs off,' and he fainted. White and

Anderson could see that the shark had taken all the flesh on Bruder's left leg above the knee and his right just below the knee. Both feet and ankles were missing and it had taken a chunk out of his side, leaving teeth marks under the arm. On the beach, women were said to have fainted or fled, according to the *New York Times*, and some 'had to be assisted to their rooms'. The two lifeguards tried to stem the flow of blood. Before the doctor arrived, however, Bruder died.

By now, the hue and cry had well and truly gone up. The Essex and Sussex Hotel telephone operator called all the major hotels and central telephone exchanges from Atlantic Highlands in the north to Point Pleasant in the south. Within fifteen minutes panicked holidaymakers were streaming from the water along 30 kilometres (20 miles) of New Jersey shoreline. Colonel William Grey Shauffler, surgeon general of the New Jersey National Guard and on the staff of the state governor, examined Bruder's body and declared that a man-eating shark had inflicted the injuries. He organised boat patrols to try to kill the big fish.

That same day, no fewer than twenty-four people died of polio in New York City, yet such was the terror that mention of the word 'shark' instilled in the general public that Bruder's death dominated the newspapers. The *New York Times* carried the headline 'Shark kills bather off Jersey beach: bites off both legs of a youth swimming beyond Spring Lake life lines: guards find him dying: women are panic-stricken as mutilated body of hotel employee is brought ashore'.

Shark fever gripped all the New Jersey holiday resorts. Boats with gun-toting vigilantes patrolled the inshore waters – sounds familiar? The cooperative Spring Lake meat markets donated prime mutton, said to have been the best bait for sharks. Local fishermen set out with huge hooks and sturdy lines to catch the culprit, but no shark was shot, caught or even seen.

The newspapers highlighted the ongoing controversy of whether sharks will attack people. One suggestion was that a giant mackerel was to blame, probably a reference to a barracuda, as Bruder's legs had been torn rather than clean cut. Local councillors played down the incident, and a ban on skimpy bathing suits that exposed 'the nether extremities' was the main talking point in Atlantic City. Even so, sharks and tales of sharks could be heard in bars and restaurants along the length of the Jersey Shore. Some brave souls swam as they normally did, and a flurry of publicity accompanied the building of a shark-proof metal fence and the introduction of a shark patrol boat at Asbury Park. Beach Haven, Spring Lake and Belmar followed suit, as did other resorts. And, suddenly, as if out of nowhere, the sharks appeared. Nobody had paid any mind before, but now any shark was considered 'rogue'.

At Spring Lake on Saturday 8 July, a lifeguard wielding an oar battered a shark swimming about 25 metres (27 yards) from the shore. Was it the shark that killed poor Bruder? At Bayonne, near Newark in the north, a group of boys swimming off a yacht club float spotted a shark. The policeman who came to their rescue emptied his revolver at the dark fin, but the shark fled to the open sea. At the Rockaway Peninsula, in the borough of Queens, New York, fishermen digging for sandworms saw a shark driving bait fish towards the shore and dispatched it with oars, spears and spades. Such was the desire for revenge that people took pot-shots at anything that looked remotely like a shark, including dolphins, porpoises and seals.

The widespread panic, though, cost New Jersey resort owners dearly, an estimated $5 million in today's money. The cancellations came streaming in, and with only six weeks left of summer holidays, they would have to work hard to claw back anything at all that season. Therefore, it was with some relief that the sobering voices of Dr Frederic Augustus Lucas, director of the American Museum of Natural History in New York, Dr John Treadwell Nichols, curator of the museum's Department of Fishes, and Dr Robert Cushman Murphy of the Brooklyn Museum announced that it was unlikely that a shark would attack anyone, but as a precaution bathers should stay close to shore. As a result, the fear subsided somewhat, even though sightings of sharks had increased enormously.

On Monday 10 July, the *New York Times* reported that the final work on the shark fence at Asbury Park had been completed and work had begun on a similar barrier at Ocean Grove but added that four sharks had been sighted off Asbury Park and another off Bridgehampton, Long Island. At about the same time, the US Bureau of Fisheries in Washington, DC, added its voice to the debate, stating that a single shark probably killed the two young men. Through hunger, the scientists speculated, it had singled out Van Sant, and having a newly acquired taste and liking for human flesh, it continued to swim close to shore and attacked Bruder. The US Commissioner for Fisheries was quick to add: 'The case is extremely unusual. I don't look for it to happen again. The fact that only two out of millions of bathers have been attacked in many years is the evidence of the rarity of such instances.'

So, the experts had spoken and again everyone was reassured, but then attention switched from the Jersey Shore to a winding tidal inlet of the sea about 50 kilometres (30 miles) north of Spring Lake on the south side of Raritan Bay, a body of water that blends with Lower Bay and the entrance to New York harbour. It's called Matawan Creek and it leads to the small town of Matawan.

Matawan is a little over 6 kilometres (4 miles) from the Atlantic Ocean and, one would have thought, immune to the threat of shark attacks. The creek was a favourite place to cool off, and in July 1916 local boys were doing just that. The body of water in which they played was extremely narrow and at low water it was not much more than a stream, but when the tide came in, bringing in shoals of small fish from the sea, they swam and fished here as they had done on almost every summer's day since anyone could remember. On this occasion, however, the incoming tide brought in something a little more sinister.

The first hint of danger was when fourteen-year-old Rensselaer Cartan, known as Rennie to his friends, was playing tag on the pilings of the dilapidated Propeller Wyckoff Dock. He leaped into the water to avoid being caught, and as his head went under momentarily, he felt something like a strip of sandpaper scrape across his stomach. He headed for the pier, where he found his abdomen streaked with blood. He called to his companions: 'Don't dive in any more – there's a shark or something in there,' but nobody paid much attention. Despite the recent events at Beach Haven and Spring Lake, life went on as usual in Matawan.

A couple of days later, during the late morning of Wednesday 12 July, the air temperature approached 32°C (90°F). Captain Thomas Cottrell, a retired seaman and part-time fisherman, left his bait-and-tackle shop in Keyport, at the mouth of Matawan Creek, for a breath of fresh air and crossed the new drawbridge that spanned the creek a few miles downstream from Wyckoff Dock. Looking down, he saw a dark grey shadow coming up the creek with the incoming tide. He shouted to two workmen, and they spotted the grey shape too. It could only be one thing – a shark. He estimated it to be about 8 feet (2.4 metres) long, but he was puzzled. What was a shark doing heading into this narrow and muddy creek that became no more than a trickle at low tide?

One of the workmen phoned John Mulsonn, the local barber and Matawan's chief of police. Cottrell first rang people in downtown Keyport to warn bathers around Raritan Bay and then ran as fast as he could to Matawan, where he tried to stop groups of boys heading to the creek to go swimming. He raced up and down Main Street shouting his warning, but people just laughed and put it down to shark hysteria. 'A shark in the creek?' they jibed. 'Pull the other one!'

The creek is just 11 metres across at its widest part, so the news did seem a bit far-fetched. Chief Mulsonn did not even bother to leave his shop. Exasperated, Captain Cottrell popped into the local dry cleaners, a new business which a strapping, blond-haired lad by the name of Watson Stanley Fisher had started, to ask for help. The business was not doing too well, so Fisher took in orders for men's suits as well. Cottrell learned that a few days previously, a man had ordered a suit and paid for it in an unusual way. Instead of cash, he offered Fisher a $10,000 life insurance policy, although what a young man of twenty-four years with his whole life in front of him would need with life insurance was a mystery to his friends and family, and they pulled his leg mercilessly. It was to be a bittersweet joke.

At nearby Anderson's Saw Mill, twelve-year-old Lester Stillwell was working alongside his father constructing wooden boxes, and at lunchtime, because the weather was so stiflingly hot, he was given the rest of the day off. He and four pals – Johnson Cartan, Frank Clowes, Albert O'Hara and Charles Van Brunt – headed for Wyckoff Dock for a bit of skinny dipping to cool off. They must have been aware of the shark scare down the coast and might even have heard Captain Cottrell's warnings, but they chose to ignore it all, as did most people in Matawan. Their ignorance did not last long.

O'Hara was climbing out of the water when something brushed past him. He looked down and saw the tail of what he later described as 'a very large fish'. Van Brunt was in the water and he saw it too, but Stillwell had disappeared. It was a shark, and it had grabbed Stillwell and dragged him below. There was blood in the water.

The other boys scrambled out and, without dressing, ran up the steep dirt road to fetch help. Workers from the nearby Fischer's bag factory came running, as did people from the town. 'A shark's got Lester!' the boys yelled. Stanley Fisher put on his swimming trunks, left his shop in the care of the errand boy and raced to the creek. At first, people thought Stillwell had had an epileptic fit, as he was prone to do, and they ran up and down the bank shouting his name.

Mary Anderson, one of Matawan's schoolteachers, shouted after Fisher. 'Remember what Captain Cottrell said this morning. It may have been a shark!'

Fisher hesitated for a moment, and then ran on. 'I'm going after that boy.'

At the creek, Fisher's father, a retired commodore with the Savannah Line, took command. Captain Cottrell led men in boats, and they poled for Stillwell's body. Others brought a roll of chicken wire, and it was placed across the creek, weighted down with stones so the tide wouldn't take out the body, but they had failed to realise that it also trapped the shark. Fisher swam to a deep spot, close to the far bank, where he thought the shark might be lurking with Stillwell. He planned to flush it out and drive it into shallow water downstream to be trapped by the chicken wire barrier.

George Burlew and Arthur Smith joined Fisher in the creek, diving down into the turbid water and feeling in the mud for the boy. Smith was close to where Fisher was diving, and suddenly felt the shark brush past him, leaving blood on his skin. Seconds later, Fisher himself appeared in a churning foam of red water.

'He fought the fish like a madman,' Burlew later told journalist and adventurer Floyd Gibbons, 'striking and kicking it with all his might. Three or four times during the struggle the shark pulled him under, but each time he managed to get back to the surface. He seemed to be holding his own, but at best, it was an uneven battle. The shark was at home in the water, Stanley wasn't.'

Eventually, Fisher wrenched himself free and stood unsteadily waist deep with his back to the people on the opposite bank. Arthur S. Van Buskirk, local deputy of the Monmouth County Detectives Office, and another man, W. H. Byrne Jr, reached him in a boat and could see Fisher holding the remnants of his right leg in both hands. As he fell forwards Van Buskirk grabbed him, dragged him over the prow and carried him back to the dock where many gentle hands carried him first to the railway station, where local physician Dr Reynolds attended to him as best he could, and then on by train to Monmouth Memorial Hospital in Long Branch, but it was too late: he succumbed to his devastating injuries as he was about to enter the operating theatre. Before he died, however, Fisher said that he had found Stillwell's body, had seized it from the shark, and thought it his duty to recover it, even though his own life was at stake. Dr Reynolds recalled his last words in an interview with the *Newark Evening News*.

'I knew it was all up with me when I felt his grip on my thigh,' said the dying young man. 'It was an awful feeling. I can't explain it. Anyhow, I did my duty.'

Fisher died at 7.30 that evening.

In Matawan, several men went to Asher P. Wooley's general store and bought dynamite. They were going to blow up the shark, which they felt sure was still close to Wyckoff Dock. The creek was cleared of boats but just as the charges were to be set off, a motorboat appeared from downriver. Lawyer Jacob R. Lefferts was at the helm and lying in the bottom of the boat was a boy, his right leg wrapped in bloody bandages. 'A shark got him,' Lefferts shouted.

The boy was ushered into a car and driven at speed to St Peter's Hospital in New Brunswick. He would not give his name at first, in case his mother was angry with him, but eventually they discovered he was fourteen-year-old Joseph Dunn. He had been swimming with three other boys from Cliffwood at the dock of the New Jersey Clay Company brickyards, about half a mile (800 metres) down the creek from Matawan. They were all in the water when news of the shark at Wyckoff Dock reached them. They quickly scrambled out, but Dunn was last. As he climbed the ladder, the shark grabbed his leg, 'like a great pair of scissors', he said later, but he did not succumb to the attack. Joseph kicked out with his free leg, while his older brother Michael and two others grabbed him and pulled, trying to drag him free, but in the tug-of-war the shark's teeth ripped the flesh off his leg. 'I felt my leg going down the shark's throat', he recalled, 'and I believe it would have swallowed me.'

Eventually, the shark did let go and Dunn was hauled to safety. He may have had a severely damaged leg, but he was alive.

Joseph Dunn was the third victim in less than an hour.

Throughout the night and the morning of the next day, townsfolk wreaked havoc in the creek. The sound of dynamite exploding could be heard from miles away, as could the shots from pistols, rifles and shotguns. Hundreds of men came from all over the district to help catch the shark. They wielded boat hooks, pitchforks, scythes and spades, and even took down old harpoons from over mantelshelves. At low tide they waded into the mud armed with knives and hammers and criss-crossed the creek with a tangle of fishing nets and more chicken wire. This town wanted vengeance, and it was determined to get it.

Then, the newspapers arrived. One tabloid organised a shark hunt, hiring a boat and filling it with men sporting rifles. Exceptionally large quantities of dynamite were let off for the benefit of newsreel cameras. Local stores ran out of explosives and ammunition. They sent to Perth Amboy on the northern side of the Raritan River estuary for more. It was as if the Great War had come to Matawan.

As the tide came in, so did reports of sharks being trapped in the creek, and when the tide went out there followed a spate of sightings of sharks escaping to the sea. The *New York Times* reported: 'Many see sharks, but all get away: Matawan's population, with weapons and dynamite, seek man-eater that killed two'. It also reported that bathing had 'almost come to a stop along the Jersey coast' and a new sport had sprung up – hunting sharks.

Scientists who had previously gone public to say that sharks did not attack people had to change their view. They blamed the attacks on a single rogue shark that had strayed north of its normal warm water habitat. In fact, everyone who had an opinion was sure to voice it. Local fishermen thought the attacks were due to ordinary sharks turning ravenous by a lack of fish on which to feed, but others declared that fish were plentiful that summer and sharks were coming inshore to catch them. In letters to newspapers, all manner of bizarre scenarios were imagined.

Even the war was blamed. In one letter, it was suggested that the sharks had followed German U-boats to the east coast of the USA. Another drew attention to sharks that followed ships entering and leaving New York Harbour. Before the war, crews would throw refuse over the side, but austerity brought on by the war in Europe meant that sharks had to find a new food supply. One *Times* letter writer even suggested the attacker was a sea turtle.

At the coast, meanwhile, there were more scares. The first was at Sheepshead Bay on the southern shore of Brooklyn, New York. Holidaymakers were having breakfast on the veranda of the Beau Rivage Hotel when they spotted a fin close to Thomas Richard, the hotel's assistant steward, who was swimming from a motorboat in front of the hotel. Several people screamed. The man raced for the boat and clambered aboard, drawing in his legs a fraction of a second before a ripple passed where he had been. The second was at Coney Island, where dancer Gertrude Hoffman had the presence of mind to splash and beat the water when she saw a fin close to her. She was not sure whether this was all for nothing or that she had barely escaped death, she said later. In the same edition of the *New York Times* that reported the near misses, there was also a reference to two large sharks in the Hudson River, and a 15-foot (4.6-metre)-long shark 10 miles (16 kilometres) off Sandy Hook on the north-east corner of New Jersey, something that would not have commanded even an inch of space in newspapers normally.

At Matawan, two days after his death, Lester Stillwell's mutilated body surfaced about 250 metres (270 yards) upstream from where he had died, dislodged probably by a storm the previous night. And so, the only lull in the frantic goings on up and down the creek was when Matawan buried its dead. Stanley Fisher's parents used the money from his life insurance policy to place a stained-glass window depicting the town of Bethlehem in the Methodist Episcopal Church on Main Street. He and Stillwell were buried at Rose Hill Cemetery. Meanwhile, a talented surgeon, Dr Faulkingham, at St

Peter's Hospital tended Joseph Dunn. He stitched and sutured and saved Dunn's shredded leg. The lad walked from the hospital exactly fifty-nine days after he had been carried in.

Six days after the attacks, however, a shark was caught in Matawan Creek. The fisherman was none other than Captain Cottrell. He was in his boat *Scud* with his son-in-law when they saw the tell-tale fin of a shark cutting through the water. They were close to the mouth of the creek, where it enters Raritan Bay, and not far from where Cottrell had first spotted the dark and menacing shadow. Setting gill nets, they trapped the shark and, after a struggle, hauled it ashore. It was 2 metres (6.6 feet) long and weighed 104 kilograms (230 pounds), but the species went unrecorded, the local *Matawan Journal* reporting that it was a 'blue-nose' or 'diamond-tooth' shark, names that are not in general use today, but could refer to the blue shark. It is a relatively harmless, unaggressive but inquisitive species that will circle, say, spear fishermen with their catch and might come in to make a test bite. Even so, although it has racked up at least thirteen shark bite incidents in the International Shark Attack listings, it is unlikely to have attacked with the ferocity that had left people with shredded limbs or even killed them.

Cottrell's net had been damaged by the shark, so he placed the shark on ice and put it on display, charging ten cents a person to take a peek at the 'Terror of Matawan Creek'. More than 300 people came to gawp, but privately he did not think it was the shark that killed Stillwell and Fisher and severely injured Dunn. However, it transpired that there was another shark caught just two days after the attacks and it became the number one suspect.

New York lion tamer and taxidermist Michael Schleisser

had joined the vigilantes intent on catching the rogue shark and, on Friday 14 July, he was dragging a net behind his boat in Raritan Bay, a little to the north of the mouth of Matawan Creek, when he snagged a shark. It was about 2.3 metres (7.5 feet) long, weighed about 147 kilograms (324 pounds), and it fought like fury. Schleisser clubbed it to death with a broken oar, before it sank his boat, and brought it back to the dock at South Amboy. When the stomach was ripped open, he discovered 7 kilograms (15 pounds) of human remains inside, including a 28-centimetre (11-inch)-long shin bone and part of a human rib. The shark had undoubtedly eaten at least one person, and the size of the shin bone, if intact, indicated that it could have been from a twelve-year-old child. The villain and its macabre stomach contents were identified by John Tread-well Nichols, the curator of the Department of Fishes at the American Museum of Natural History mentioned above. The shark, he confirmed, was a white shark, *Carcharodon carcharias*, but was it responsible for all these attacks?

At the time, the general feeling had been that a shark – any shark – was unlikely to attack a person in America's inshore waters, especially in more northerly locations. One business-man even offered a reward for proof of a shark attack north of Cape Hatteras, North Carolina, and there were no takers. And, even if a shark did take a bite out of somebody, its jaws were not strong enough to sever a leg, said at least three lead-ing scientists of the day. The summer of 1916 changed all that.

By mid-July that year the shark bite incidents mercifully ceased, even though it was what locals called a 'shark year', with the appearance of many sharks of several species off the east coast. However, this was when the speculation first started, and surprisingly has continued to this day. First up were Nichols and Robert Murphy (from the Brooklyn Museum, you'll recall), who wrote in the *Magazine Section* of the *New York Times* on 6 August 1916:

> White sharks are so scarce that their habits are little known …
> Of this species it may be said that, judging from its physical
> make-up, it would not hesitate to attack a man in the water
> … Even a relatively small white shark, weighing two or three
> hundred pounds, might readily snap the largest human bones
> by a jerk of its body after it had bitten through the flesh. The
> occurrence of the white shark near New York being almost
> unprecedented as were the attacks on bathers which happened
> simultaneously, the capture of a specimen by Mr Schleisser
> confirms our belief that the white shark was responsible for
> the casualties.

Hugh M. Smith, United States Commissioner of Fisheries, was more cautious in apportioning blame, and in the notes section of a paper published in the *American Museum Journal* in October 1916 he took Nichols and Murphy to task. He wrote: 'It is interesting to quote the opinion of Mr J. T. Nichols of the American Museum and Mr Robert C. Murphy of the Brooklyn Museum, who consider the circumstantial evidence sufficient to convict the white shark in spite of the lack of definite proof against it.'

In his own submission, he stated:

> The Bureau of Fisheries was incessantly importuned to explain
> why sharks were behaving as they were, and to take action that

would prevent further attacks. There was some criticism of our inability to cope with the situation, although obviously there was little that could be done. The culprits were never identified. It was not known whether one individual shark of a species common to the region was running amuck; whether representatives of several local species had been forced to attack human beings because of certain undetermined biological or physical conditions; or whether there was an advent of shark or sharks from distant waters with feeding habits different from those of the domestic species, which in no former years had exhibited any man-eating tendency and were dangerous only when they themselves were attacked.

So, in 1916, despite all the publicity and analysis, the question was still begging, and it's still begging now: was Michael Schleisser's white shark really the shark that attacked five people, killing four and severely injuring one? In recent years, scientists have deliberated endlessly on the pros and cons and have felt that it was unlikely. It was thought generally that white sharks are more oceanic in their water preference and rarely entered rivers, although juveniles do enter saline bays and larger estuaries.

Young white sharks, for example, are occasional visitors to Australia's Lake Macquarie and Lake Illawarra in New South Wales, not actual lakes but coastal lagoons connected to the sea. They follow fish shoals that move in from the ocean. In March 2014 a juvenile white found itself high and dry on the rocks after chasing small fish near the ferry that crosses the Crookhaven River, joining the mainland with Comerong Island, south of Sydney. The ferryman and a group of local anglers revived the young shark and sent it on its way. And, on the odd occasion, bigger sharks have been spotted. In May 1987, for example, a 4-metre (13-foot)-long white shark ventured into Tallebudgera Creek on Queensland's Gold Coast, and was speared for its trouble, but how far upstream it travelled is not reported.

Other than these youngsters and a few adults, it was thought that white sharks avoid fresh or brackish water. More usually, this environment is the domain of two other types of shark – the rare river sharks of the *Glyphis* genus, which can be ruled out because they frequent rivers and estuaries in northern Australia and Papua New Guinea, and the notorious bull shark *Carcharhinus leucas*, which has a much wider, almost global, distribution in tropical, subtropical and warm temperate waters. The bull shark, more than any other species, is likely to be encountered in a muddy creek on the east coast of the USA, and it is even more aggressive than its white shark cousin. It's a particularly dangerous animal and can be found as far north as Massachusetts in summer.

Bull sharks regularly enter fresh water. It's a survival tactic. Females head up rivers and into lakes, including the remote Lake Nicaragua in Central America, to drop their pups, safe from the cannibalistic attentions of their fellow sharks. They have been found 4,000 kilometres (2,500 miles) up the Amazon and as far upstream as Arton, Illinois, on the Mississippi. In Queensland, Australia, they have been seen swimming up one of the main streets in Brisbane during floods and have been trapped in a pond at a golf course on the city's outskirts. They enter the Ganges and Brahmaputra rivers in India and Bangladesh, where they feast on the human bodies that have been partially cremated, and the St Lucia estuary in South Africa and Kipling's great, greasy Limpopo, where they attack hippos as well as people, and they have even been seen in the Potomac River, which flows through Washington, DC, some way to the south of the Jersey Shore.

Wherever they exist they are potential man-eaters. They are the bruisers of the shark world, the bullyboys that have no fear. In the murky waters of an estuary, they are in their element, so Matawan Creek would have been prime bull shark territory, the place where you'd expect to find them.

The behaviour of the shark towards Stanley Fisher is also not an obvious trait of a white shark. The white shark is a sneaky stalker, a killer that relies on surprise and avoids confrontation. Stanley was faced with the ocean's equivalent of a leopard, an aggressive and fearless shark that was prepared to fight for the prey it had acquired. With this in mind, the Matawan Creek monster is less likely to have been a white shark and more probably a bull shark. The attacks off Beach Haven and Lake Springs, however, are a different story.

Here, the swimmers had entered the province of the white shark and, as these animals migrate regularly, north and south depending on the season, along the east coast of North America, for a swimmer to encounter one off a New Jersey beach is not surprising, especially a swimmer who was outside the normal melee of beach activity as both Charles Van Sant and Charles Bruder had been. White sharks, after all, rely on surprise to take their prey. They avoid the chaos of a busy beach. The shark that Michael Schleisser had caught might have taken a bite out of at least one of these two lads, if not both, for they were remote from the other bathers.

So, it was generally felt in scientific circles that although the nearshore attacks were the work of at least one white shark and maybe two, the Matawan Creek attacks were more likely to have been by one or more bull sharks; and there the story remained until November 2013.

Fourteen months previously, in September 2012, researchers from OCEARCH placed a GPS satellite tag on Mary-Lee, a 16-foot (4.9-metre)-long female, weighing 3,500 pounds (1,600 kilograms), the weight of a medium-sized car. The tag was bolted to her dorsal fin, and every time her fin cut above the surface, it transmitted information about her location to a satellite overhead. This way scientists could follow her progress in real time across the ocean. She was tagged off Cape Cod, Massachusetts, and when released, she immediately went on her travels. She swam along the east coast of the USA in a series of dives between the surface and a depth of 40 metres (130 feet), followed by a sojourn in deep water to Bermuda, an excursion south to the Bahamas, and a visit to the surf line off the Florida coast, where she probably scared the heck out of local surfers, but in November 2013 she surprised everyone by heading into St Helena Sound, a coastal inlet with shallow marshes in South Carolina. The brackish water there at high tide, according to local shrimper Don Anderson, is about 20–25 feet (6–7.5 metres) deep and the estuary is filled with shrimp and baitfish, as well as redfish and speckled trout… a place somewhat similar to Matawan Creek.

Scientists began to reassess the evidence. Aside from the new Mary-Lee data, a Smithsonian Channel TV show, *The Real Story: Jaws*, revealed that on the day of the Matawan Creek attacks the full moon and a spring tide would have doubled water salinity a couple of hours before high water, so any shark entering the creek would be riding in on a wedge of high-salinity seawater. Conditions would have been more like the open sea than a fetid creek. In addition, George Burgess and his team at the International Shark Attack File discovered the Matawan inlet to be more a marine embayment than a brackish water creek, big enough for tugboats and barges to operate in, and so it could easily have hosted, at least temporarily – as long as there was food to be had – a young, inquisitive and bold white shark. It means that, back in July 1916, a white shark could well have entered Matawan Creek just as Mary-Lee had done in South Carolina in November 2013. Similarly, in 2022, a juvenile white shark entered the relatively shallow water of a tidal estuary on Cape Cod, and there's a very clear video of the event on Facebook, so this is yet more evidence pointing to the Matawan culprit probably being a young white shark.

The white shark continues to surprise us, even now… but there's more. Mary-Lee is also implicated in solving the identity of a shark that attacked a young man in St George's Creek in St Mary's County, Maryland. The 'creek' is actually a large inlet that opens near the mouth of the Potomac River which, in turn, opens into Chesapeake Bay, and the circumstances of the shark attack were not unlike those at Matawan. The year was 1640, and the summer had been especially hot. People in seventeenth-century garb were roasting in the August sun, so an unnamed Englishman and his friends took off their clothes and went swimming to cool off. The young man initially went in alone, when he was suddenly seized by what was described as 'a huge fish'. It was the first known shark attack in historical times in North America, and shark aficionados have since tried to work out what species was the offender. Like the Matawan incidents, it looked as though a bull shark was responsible. It was the most obvious candidate as bull sharks frequent the area in and around Chesapeake Bay during summer, and they are known to attack people. A largish specimen, up to 3.5 metres (11.5 feet) long, could do a lot of damage. However, like Matawan, white sharks turn up here too.

Chesapeake Bay is not a natural habitat for them, but they are occasional visitors. They've been known to enter the bay in the vicinity of Virginia Beach, which is on the Atlantic coast but close to the southern shore at the bay's broad entrance. In June 2018, two were spotted right here just days apart. They were observed during a survey by the Virginia Institute of Marine Science. The first was an 8.5-foot (2.6-metre)-long male off Virginia Beach and the other a 12–13-foot (3.7–4.0-metre)-long white shark about 4 miles (6 kilometres) offshore from Sandbridge, just to the south, also on the Atlantic coast.

There have been more. In November 2020, a 17-foot (5-metre)-long female white shark being tracked by OCEARCH, which scientists nicknamed Nukumi, was in Maryland waters. In June 2017, Mary-Lee was 20 miles (32 kilometres) off the Maryland coast. In July the same year, a 12.5-foot (3.8-metre)-long male white shark called Hilton passed close to the mouth of the bay, and in October the same year, another large male, about 13 feet (4 metres) long and named Bob, did the same.

The sharks are not residents but generally pass by during their migrations north and south along the Atlantic coast. Like Mary-Lee in South Carolina, some stop off briefly, and may enter bay waters so, although the likely identity for the St George's Creek attacker is a bull shark, there is an outside chance it was a visiting white shark just like in the Matawan Creek incident 276 years later.

PART II

ANCIENT ROOTS

CHAPTER 5

460-MILLION-YEAR PREHISTORY

Until the seventeenth century, nobody knew what fossils were, but that they existed was known from prehistoric times. Early people dug fossilised sharks' teeth out of rocks or found them lying about on the ground and used them as cutting tools, but what they represented and where they came from was a mystery to them. In the first century CE, the Roman philosopher and naturalist Pliny the Elder put forward the notion that the triangular objects rained down from space during solar eclipses, but much later, during medieval times, fossilised sharks' teeth were thought to be petrified dragon's or serpent's tongues, known as *glossopetrae* or 'tongue-stones', and they were supposed to have magical medical properties.

They acquired their supposed miracle cures care of St Paul. It was said that he was shipwrecked on the island of Malta and, when he came ashore, he was bitten by an adder. He shook off the snake and placed a curse on all the snakes living on the island, turning their forked tongues to stone.

Medieval apothecaries placed the fossil teeth on the bite mark made by a venomous snake in order to save the victim's life, and they were dunked in a chalice of wine to neutralise any poisons that had been put in there by the drinker's enemies.

They were often mounted in silver and so made into a pendant worn around the neck, with the belief that they would help to protect the wearer. They were also ground into powder and given as a remedy for epilepsy, plague, poxes, fevers and bad breath. Needless to say, they didn't work, but it didn't stop them from changing hands for significant sums of money, teeth from Malta commanding the best prices.

Even though magical tongue-stones were used as amulets right up until the early twentieth century, and in parts of rural Britain were known as 'cramp-stones' to ward off cramps, rheumatism and toothache, in 1666, things began to change. That was the year that French fishermen caught a white shark off the Tuscan coast and its head was transported to Florence and dissected by Danish anatomist Nicolaus Steno (see also Chapter 2). During the course of the dissection Steno noticed that the white shark's teeth looked exactly like the tongue-stones that people collected on the hills around Tuscany. He went on to propose that the origin of the tongue-stones was teeth from sharks that lived during a previous age. It was a groundbreaking discovery and it led to the modern-day scientific discipline of palaeontology, in which scientists could reveal what life was like millions of years ago.

White sharks, in fact, have a long and distant ancestry, but teasing it out has not been easy for palaeontologists. For one thing, fossils of sharks are rare, as their cartilaginous skeletons do not preserve well. The only hard parts are teeth and dermal denticles or skin teeth, scales that resemble miniature teeth embedded in the skin (see also Chapter 6). Down the ages, there are trillions left behind by sharks, but little else. So, it's remarkable that palaeontologists can reveal as much as they do about ancient sharks from just a tooth or a scale.

The very oldest specimens are, indeed, little more than tiny scales, many less than 0.4 millimetres (0.02 inches) across and found in rocks from the Ordovician period, 460 million years old. Scientists are not at all sure that they came from the earliest true sharks or simply from shark-like animals. Whatever they are, each distinctive scale has given rise to a new species name. Nobody knows what they looked like or how they lived, and possibly we never will.

So, in 2003, when a fossil shark skeleton was discovered at a site in New Brunswick, Canada, it was greeted with great excitement. Scientists named it *Doliodus problematicus*, meaning 'problematic deceiver'. They estimated it to be about 409 million years old, from the early Devonian period, making it the oldest intact shark fossil ever found. When alive it would have resembled the modern bottom-dwelling angel shark, except for a long spine on each pectoral fin. It also had replaceable rows of teeth, a feature of modern species that was thought to be absent in the earliest sharks.

By the middle of the Devonian – the so-called Age of Fishes – fish that more resembled torpedo-shaped sharks began to appear in the fossil record. One of the first was found in rocks in the Antarctic and Australia, which gave rise to its scientific name, *Antarctilamna*. It lived about 380 million years ago and had a vague resemblance to a modern bullhead shark with a substantial spine in front of the first dorsal fin, except it had more primitive features. It had a slightly more elongated eel-like body, a symmetrical tapering tail and a mouth at the front of its face rather than underslung like modern sharks, and in its jaws each of its teeth had a two-pronged crown.

These early sharks were evolving at an interesting time in earth's history. The surface of the earth was dominated by two giant continents – Euramerica in the north, with a bit of Siberia to the north again, and Gondwana in the south. They were surrounded by warm shallow seas that spawned a huge diversity of lifeforms: enormous sea scorpions up to 2 metres (6.6 feet) long, hundreds of species of armoured and jawless fishes, along with a monster of a fish – *Dunkleosteus*, a giant placoderm or 'plate-skinned' fish up to 10 metres (33 feet) long that chomped on ammonites and other nautiloids, armoured fishes and early sharks with a bite that is thought to have had the greatest force of any fish living or extinct. It could also open and close its mouth in fifty to sixty milliseconds, so prey was sucked into its mouth. It was one of the first apex predators to roam the seas, a position several species of sharks, including white and tiger sharks, hold today, and even though the early sharks played their role in the Devonian ecosystem, they were not in the same league as *Dunkleosteus*.

At about the same time lived *Cladoselache*, a more shark-like fish about a metre (3.3 feet) long. Nobody is sure whether it was a shark or not, but it certainly looked like one with its torpedo-shaped body, short, stout spines in front of each of its two dorsal fins, large paired pectoral fins and a crescent-shaped tail fin. It had a long, slender mouth positioned terminally at the front of the head. The top and bottom lobes of the tail were about equal, a similar arrangement to the white, porbeagle and mako sharks. That's indicative of a fast swimmer, so it's thought *Cladoselache* was a high-speed predator, a bit like a modern mako shark, the world's fastest living shark.

So, *Cladoselache* might have been fast, but its swimming was not helped by the near-absence of dermal denticles. As sharks go, it was almost naked. Unlike a modern shark that has these tooth-like scales all over its body that help it move effortlessly through the water, *Cladoselache* just had a few scales on the edge of its fins, in the mouth and around the eyes – not really much help in swimming hydrodynamics. Maybe the anatomical setback meant that it occasionally came a cropper to that monster *Dunkleosteus*. Another surprising absence was no claspers – the pair of penis-like structures that hang loose close to the vent of most other species of sharks. It could be that nobody has found them yet, or the species indulges in external fertilisation. Whichever way, it's very odd.

At the end of the Devonian period, the planet was hit by a series of mass extinction events, probably due to changes in the climate, when about 75 per cent of all the species on Earth were killed off. The placoderms were very hard hit, so there were few armour-plated fish surviving. This left an ecological hole which sharks helped to fill. With the competition gone, sharks were able to take over ecological niches left vacant by the extinction victims. As a consequence, sharks proliferated and diversified in an extraordinary way.

During the Carboniferous period, which started about 359 million years ago, the two continents were amalgamating into a single supercontinent, which we call Pangaea. Part of the supercontinent wrapped around a sea known as the Paleo-Tethys Ocean and a vast super-ocean, called the Panthalassic

Ocean, surrounded the entire new continent and dominated the surface of the planet. It was the Age of Amphibians on land, where the 'coal forests' thrived, but in the ocean sharks and their relatives dominated for the first time. There was a so-called 'golden age of sharks', when some took on very strange appearances. There was, for example, a shark-shaped fish – *Stethacanthus* – that had what resembled a flattened shaving brush where its anterior dorsal fin should be, and another – *Helicoprion* – with a whorl of teeth in its lower jaw that looked like a circular saw. Then, there was *Falcatus*, the males of which had a strange, thick spine that stretched forwards over the top of the head.

The scissor-toothed sharks, such as *Edestus giganteus*, were large, up to 6 metres (20 feet) long, and shark-like, but their jaws and teeth were very strange. The jaws seemed to have been splayed apart and the teeth stuck out so they resembled pinking shears, but in some reconstructions, the upper and lower jaws do not seem to meet. How they caught anything, therefore, was pure speculation, although it had been suggested that they slashed at prey to disable it and then returned to suck it up; but in 2019, researchers at the Idaho Museum of Natural History published a paper that revealed what they thought was going on.

One of the things they found is that the jaw mechanism has two unexpected joints. During a bite, the upper and lower teeth whorls move towards each other, with the bottom layer moving backwards and forwards, to and from the throat. It serves to slice the prey in half, like slicing a piece of bread. It's a unique way of feeding which hasn't been seen before. Leif Tapanila, curator of the museum and lead researcher, who also researched *Helicoprion*, told *Smithsonian Magazine*: 'The jaw of *Edestus* seems to be primarily developed to quickly and fatally amputate its prey, snapping them in half, and presumably picking up the falling pieces after the victim was subdued.' Evidence for this has been decapitated and de-finned chunks of fossil fish found in the same fossil beds as the shark. Why these sharks, and *Helicoprion*, developed such unusual biting mechanisms is still a bit of a mystery, although there's the suggestion that the sudden abundance of 'meaty' prey, such as squid and other fish, during the Carboniferous period might well have triggered the evolution of sharks that slice.

And if that wasn't enough, the so-called 'Godzilla shark', which lived about 300 million years ago, was another with unusual teeth. Each tooth was about 3 centimetres (1.2 inches) across with five spear-like cusps, and there were twelve rows of them in the lower jaw. They would have been used to grab and crush prey that lived on or near the seabed. About 2 metres (6.6 feet) long, the shark's most obvious physical characteristic was a large, 57-centimetre (22-inch)-long spine in front of the anterior dorsal fin which recurved back over the body. The posterior dorsal fin had a smaller but still formidable spine, so the fish was well protected against predators. It was not a true shark, but a shark-like ctenocanth, a group that had larger but less flexible terminal mouths than sharks. Godzilla, or more correctly *Dracopristis hoffmanorum* or Hoffman's dragon shark, was named in honour of the New Mexico family who owned the land on which the fossils were found.

Another oddball tooth-wise was *Petalodus*. Its fossil remains were found in limestone rocks about 290 million years old, towards the end of the golden age during the Permian period.

It was an active shark-like predator that more than likely migrated across the ancient oceans, but it also had unusual teeth. They were shaped like mushrooms, with a petal-shaped crown supported by a tongue-like shaft.

These are just a few of the unusual sharks or shark-like fish that were the result of the rapid radiation of sharks after the Devonian mass extinction; in fact, during the golden age, there were thought to be more species of sharks than all the other fish species present in the oceans. Many, though, were nature's flamboyant but failed experiments, some most probably from the chimaera lineage rather than true sharks. Chimaeras are relatives of sharks whose line split from the sharks about 420 million years ago and today occur exclusively in the deep sea.

Those sharks which lived on and evolved into fish that we could better identify as sharks were better equipped to deal with earth's changing conditions, and without the gimmicks. The basic shark design has been remarkably durable, and these were the survivors that were able to procreate and continue their line. The quirky sharks dropped out. Indeed, sharks, as a group, have survived at least five mass extinction events during the course of earth's history. They were and still are the toughies.

At each extinction event, millions of species of plants and animals simply disappeared off the face of the earth, the most serious event occurring at the end of the Permian when almost all life became extinct due to widespread volcanic activity. Scientists have called it the 'time of great dying', because only a very small percentage of animal species lived on, including our own ancestors, and, of course, the ancestors of modern sharks. Again, the mass extinction left many ecological holes to fill, and again surviving sharks were quick to fill them.

The next major shark event in our story is towards the beginning of the Jurassic period, about 195 million years ago, when the first group of modern sharks evolved. They were, and still are, the hexanchidae, which includes the extant cow sharks and frilled sharks. The frilled sharks, also known as lizard sharks, have their characteristic frilly gill slits, and a host of primitive features, such as a 2-metre (6.6-foot)-long eel-like body with dorsal, pelvic and anal fins close to the tail. They live in the deep sea.

Like other sharks in the group, the frilled shark has six pairs of gill slits rather than the five seen more commonly on less primitive sharks. Similarly, the cow sharks have either six or seven pairs and are considered to be the most primitive of living sharks. Their skeletons are remarkably similar to those of sharks that lived during the Jurassic. They have primitive comb-like teeth in the lower jaw, and their digestive and excretory systems are unspecialised. A fossilised hexanchidae tooth, found in Devonian rocks in Japan, suggests this family of sharks is a direct survivor from the Permian-Triassic mass extinction event.

Later in the Jurassic, other groups of modern sharks appeared. Many had one thing in common: protrusible jaws that could push the mouth and teeth beyond the snout, so the shark didn't have to turn on its side or back to take a bite, as the ancient Greeks and Romans thought. One of those sharks – *Palaeocarcharias* – is thought to be the oldest known ancestor of the white shark. It was nothing like the streamlined super-predators that followed but resembled more a sluggish carpet shark, about 1 metre (3.3 feet) long, that fed on or near the shallow seafloor. We know this detail because much of the skeleton had been fossilised and located in the famous Solnhofen Plattenkalk of southern Germany, the beds in which *Archaeopteryx* was also discovered. They are unique in the palaeontological world in that the soft parts are often fossilised as well as the hard bits. The bed in which the shark lies is about 165 million years old, that is Middle Jurassic. However, the link between this unlikely forebear and the white shark is the detailed structure of its teeth.

The composition is not unlike human teeth: a hard shell of enamel (humans) or enameloid (sharks) with a dentine core. The dentine can be of two types: orthodentine, which has a compact appearance, and osteodentine, which is spongy. The osteodentine is generally in the tooth's root, anchoring it to the jaw, while the orthodentine is in the crown. However, when an international research team led by scientists from the University of Vienna came to examine the dentine of the living white shark and its close relatives, they found that the osteodentine from the roots intrudes into the crown, replacing the orthodentine. It's an arrangement that's peculiar to this group of sharks. Now, what came as a surprise is that *Palaeocarcharias* teeth have the same structure, leading them to suggest that this species is the most ancient known ancestor of the white shark.

Whether *Palaeocarcharias* lived on into the Cretaceous is not known, but things were generally going well for sharks during this period. There were hundreds of species from many families, and they were holding their own among the giant marine reptiles – ichthyosaurs, plesiosaurs, pliosaurs and mo-sasaurs – that ostensibly ruled the oceans.

One way to avoid the reptiles was to hunt closer to the seabed and many species of sharks did just that, but about 93 million years ago they were forced up towards the surface. There was a period of global warming, much like that happening in the world today. At its peak, the average global sea surface temperature was 28°C (82°F), the highest it had been for 200 million years. The cause was a massive outpouring of molten lava along with huge quantities of volcanic gases, including the greenhouse gas carbon dioxide. Global temperatures rose and oxygen levels deep down in the ocean dropped alarmingly, forcing some sharks to seek higher oxygen levels closer to the surface. They had, in effect, become pelagic sharks and as such had to compete with the fast-swimming predators that occupied the open ocean. The result was that, over time, they evolved longer and narrower pectoral fins, and the warmer surface waters enabled their swim muscles to work more effectively. This gave rise to the direct ancestors of blue, oceanic whitetip and mako sharks – today's open ocean specialists – but also included the forebears of the modern white shark.

In fact, the apex predator at this transitional time was an analogue of today's white shark, an open ocean cruiser known as the Ginsu shark, *Cretoxyrhina mantelli*, on account of the shape of its razor-like teeth. They resemble a Ginsu knife for 'slicing and dicing' and can be up to 8 centimetres (3 inches) long. The shark itself had a maximum length of about 8 metres (26 feet), slightly bigger than a modern white shark, but with an appearance and build like its modern equivalent although they are not closely related. It was one of the largest species of sharks at the time, and it had a fair turn of speed, including bursts of up to 70 km/h (43 mph). It lunged at its prey at high speed, similar to the white shark today, and probably had similarly good eyesight. It fed on the smaller mosasaurs, plesiosaurs, other sharks and bony fish, and even pterosaurs and dinosaurs if they should enter the ocean. Its demise was probably due to competition with the larger mosasaurs, such as the 14-metre (46-foot)-long *Tylosaurus* with which it shared the Western Interior Seaway that split what is now the North American continent.

Another species living at this time was *Ptychodus*, a classic case of fossil teeth in search of a body, but early in 2024 scientists published a paper that began to piece together the shape of that body. It was up to 9.7 metres (32 feet) long, so a substantial shark, longer than a modern white shark. It was thought to be a fast swimmer resembling a modern-day porbeagle or salmon shark, except for one thing: rather than possessing the fish-grabbing needle-sharp teeth of the porbeagle, it had huge flat, grinding, rectangular teeth, up to 55 centimetres (22 inches) across. It's thought that it cracked open ammonites and crushed sea turtles. However, other sharks and marine creatures had the same idea and eventually the competition became too much. The species became extinct about 76 million years ago. Roughly 10 million years later many other species followed suit for by the end of the Cretaceous period, life on earth was dealt another fatal blow.

About 65 million years ago, a large asteroid slammed into the Yucatan Peninsula in present-day Mexico and the resulting Cretaceous-Paleogene mass extinction event put paid to 50 per cent of plant and animal species, including all the non-avian dinosaurs and their marine relatives. Again, the sharks came through, although not totally unscathed. The larger ones went the way of the dodo, the survivors being smaller and living in the depths of the sea.

One of those survivors was *Cretolamna*, thought to be the common ancestor of all the lamnid sharks, including the white shark, porbeagle, mako and the giant sharks of the Miocene. Its maximum length was about 3 metres (10 feet), and it resembled a porbeagle or salmon shark in appearance. It was, it is thought, a shark of the open ocean and its teeth indicate it had a generalist diet – bony fishes, sea turtles, squid, other sharks and even mosasaurs before the mass extinction whipped them away.

After the extinction event there were ecological niches to be won once more, and *Cretolamna* would have filled one. However, throughout the following millennia, some species of sharks were so successful in the evolutionary stakes that they began to evolve to an exceptionally large size.

It was about this time, during the early Eocene about 60 million years ago, that *Otodus obliquus* appeared. Like *Ptychodus*, it was one of the mackerel sharks, an informal grouping that includes the modern white, mako and porbeagle sharks, and was thought to have a maximum length of about 9 metres (30 feet), so undoubtedly another formidable beast. Reconstructions show its jaws were lined by multiple rows of triangular teeth, 4 centimetres (1.5 inches) long, with smooth cutting edges. Quite a mouthful. Food was anything from large fish and sea turtles to whales and dolphins.

It was one of a series of several giant *Otodus* species that appeared during the early Miocene period that followed, with different species evolving progressively bigger over time. One, for instance, reached a length of 13.5 metres (44 feet); but the species in this genus that really did scare the world – not the modern world, but the oceans of 23 to 3.6 million years ago – was, of course, the infamous 'megalodon', *Otodus megalodon* to give the shark its full scientific name. This really was a bruiser, a fearsome shark that is estimated to have had a maximum length of 20 metres (66 feet) – that's about the length of a cricket pitch and bigger than the largest ever recorded specimen of a modern whale shark, which measured 18.6 metres (61 feet), the main difference being that the whale shark is a gentle filter feeder, while the megalodon was a voracious predator.

However, the megalodon was not so giant throughout its range. Researchers at DePaul University, Chicago, found that specimens in warmer waters were significantly smaller than those in cooler waters. It's an example of Bergmann's rule, a concept in biology introduced in the mid-1800s by German biologist Carl Bergmann. It explains how larger animals thrive in cooler climates because their size helps them retain heat, compared to animals with smaller bodies.

Other vital statistics have been determined with the help of 2D and 3D modelling by teams of researchers from the UK, USA, Australia, South Africa and Switzerland. One group based their model on a fossil of an intact vertebral column collected in the 1860s and languishing in a Belgian museum. They concluded that this particular megalodon was about 18 metres (59 feet) long and weighed more than 61 tonnes (67 tons). It is thought to have looked superficially like a gigantic white shark or maybe a porbeagle/salmon shark, with a 1.6-metre (5-foot)-tall dorsal fin (almost as tall as the average human) and a tail about 4 metres (13 feet) in height. It differed from the extant white shark by having substantial musculature around the shorter but bulkier 4.7-metre (15-foot)-long head. These muscles manipulated jaws that, when fully open, presented a gape of more than 1.8 metres (6 feet). Two average-sized humans could stand in there and have space to spare. Inside were rows of replaceable, triangular, serrated teeth that were huge. The largest was the size of one's hand, 18 centimetres (7 inches) tall.

The force with which it could bite is nothing short of staggering – a car-crushing 18.2 tonnes according to a computer simulation by researchers at the University of New South Wales, and all the more remarkable if you consider its skeleton is made of cartilage and not bone. By comparison, the modern white shark has a bite force of 1.8 tonnes; yet kilo for kilo, according to the scientists who carried out the simulation, the megalodon's bite was no stronger than that of a domestic cat. It's the sheer size of the animal that resulted in such a frightening bite force.

Calculations revealed that its stomach capacity was between 9,605 and 10,722 litres (2,113–2,359 gallons), so it could have gobbled down a marine mammal the size of a modern full-grown killer whale in just five neat gulps. It could also have swallowed a large white shark whole. Such meals would have enabled it to travel more than 7,000 kilometres (4,350 miles) until it had to feed again. This capacity for moving over vast distances would also have enabled it to be at particular locations almost anywhere in the ocean, making prey encounter rates high, or to visit places where food was seasonally available in large quantities.

Further estimates have shown that the Belgian shark would probably have had a cruising speed of more than 5 km/h (3 mph), compared to the living white shark, which cruises at 3 km/h (2 mph). This is a faster cruising speed than any other shark, including modern species.

This picture of the megalodon, of course, has to be pure speculation because mainly teeth and vertebrae have been fossilised and not a whole shark, but it's surprising, as we have seen, what information can be gained from them. The teeth, for instance, can tell us what kinds of prey the megalodon hunted, and it's all down to isotopes.

Any isotopes of nitrogen in an animal's body are derived from the proteins in its diet. There are two of interest: nitrogen-14 and nitrogen-15, and animals generally retain more nitrogen-15 than 14, so animals higher up the food chain accumulate more nitrogen-15. This can be detected in the teeth, and at the University of California, Merced and DePaul University in Chicago, researchers have determined that megalodon teeth have very high levels of nitrogen-15, which suggests it was a true hyper-apex predator, a 'predator of predators', if you like. Main prey animals were small to medium-sized whales, between 2.5 and 7 metres (8–23 feet) long. There were, at the time, more than twenty recognised genera, compared to just six today. On the menu, for example, was the diminutive *Piscobalaena*, a typical baleen whale shape but little more than 5 metres (16 feet) in length (consumed in three bites from computer estimates), and *Xiphiacetus*, a toothed whale sometimes described as the long-snouted dolphin, which was up to 3 metres (10 feet) long, the snout or rostrum accounting for a third of that length. Evidence comes in the form of bite marks on the fossil skeletons of small whales and large seals, suggesting these petite whales were staple food for megalodons. Even so, individual sharks show variations, the biggest being between adults and juveniles. Megalodon youngsters must have targeted mainly seals, rather than whales.

Marine palaeontologists at the University of Zurich found a surprisingly large number of juvenile megalodon teeth at a location in Panama. During the Eocene, it must have been a megalodon nursery in warm shallow waters, where the youngsters would have been relatively safe from larger predators. Similar sites have been revealed in Spain, along with a birthing ground and nursery for one of the megalodon's close relatives, *Otodus angustidens*, a 12-metre (39-foot)-long megatooth shark that lived during the Oligocene and Miocene epochs, 33–22 million years ago, which has been uncovered in South Carolina. There is the possibility, however, that because these sites are in what were tropical or subtropical seas, we could be looking at small adult megalodons and not juveniles – Bergmann's rule.

In another aspect of reproductive biology, researchers studied not teeth, but individual vertebrae. They examined the fossilised vertebrae of a 9.2-metre (30-foot)-long megalodon, especially the growth rings, a bit like checking tree rings to determine a tree's age. They found that the shark must have died when forty-six years old, but more interestingly they worked out that it must have been about 2 metres (6.6 feet) long at birth. Now, that's quite a hefty baby. It incubated inside its mother, but to reach such a large size the unborn baby must have snacked on all the other eggs and maybe even the other embryos its mother produced, a behaviour known as intra-uterine cannibalism. Modern sand tiger shark mothers have these kinds of battles going on in their womb every day they are pregnant. One dominant baby consumes all its womb-mates. It ensures the victor gets a good start in life.

There is also some evidence from the minerals in the teeth that the megalodon's body was slightly warmer than the surrounding seawater, by about 7°C (13°F). Whether it had some kind of heat exchange mechanism in the blood system, like the modern white shark (see also Chapter 6), or whether its very size enabled it to retain heat is unknown, but the ability to keep its body warm would speed up digestion, make it more alert and enable it to go further and faster, and it could venture into colder waters where most other shark competitors could not go and therefore exploit an exclusive source of food.

All these superlatives mean that many shark biologists consider the megalodon to have been not only the world's largest ever shark, but also one of the most formidable, if not *the* most formidable, marine predator that ever lived. It was also what scientists have described as a cosmopolitan transoceanic super-predator which must have had a huge impact on marine communities, but it didn't have everything its own way. Competition came from some other pretty hefty creatures, such as the macroraptorial sperm whale *Livyatan melvillei*, its scientific name derived from the biblical Leviathan and the specific epithet from Herman Melville, the author of *Moby-Dick*. Like the megalodon, *Livyatan* was a monster, about the size of a modern bull sperm whale, up to 17.5 metres (57 feet) long, and with a mouth filled with enormous enamel-covered, peg-like teeth. There were two rows in the lower jaw and two in the upper, with the largest teeth up to 32 centimetres (13 inches) tall. They were the largest known functional teeth of any known animal, not including tusks. The fossil teeth showed signs of wear, which indicated that they sheared past each other when biting down. The raptorial whales chomped down on medium-sized baleen whales such as *Cetotherium*, a variety about 6.8 metres (22 feet) long. Although baleen whales were the main prey, they would also hunt beaked whales, dolphins, seals, sea turtles and sharks, precisely the same prey as the megalodon and filling a similar ecological niche to that occupied by killer whales today. In fact, *Livyatan* probably chased down its prey until it was tired out, just like pods of killer whales do, except this predator was so big and powerful it probably did it all on its own.

A cooling event at the end of the Miocene epoch, albeit still warmer than today, meant that the quantities of food required by super-predators were suddenly hard to come by and, by the early Pliocene, *Livyatan* eventually became extinct, along with the smaller whales on which it fed; indeed, there appears to have been a series of mass extinction events in the oceans caused by fluctuations in atmospheric carbon dioxide, and therefore also global temperature, during which all the mega-tooth-type sharks began to disappear, and they weren't alone.

Sometime earlier, about 19 million years ago during the early Miocene, there was a massive die-off of sharks in general. Researchers at Yale University and the College of the Atlantic chanced upon the event when they were looking at a survey of fish abundance over 85 million years. At the time, there were ten times as many sharks in the ocean as there are today, but suddenly 70 per cent of them simply disappeared, more than became extinct at the Cretaceous-Paleogene mass extinction. Unlike the other die-offs, though, there was no global climate calamity or ecosystem disruption, but during the event more open ocean species died than coastal sharks. As one observer remarked: 'It represents a major change in ocean ecosystems at a time that was previously thought to be unremarkable.' Nevertheless, the megatooth sharks survived this setback. Maybe they avoided the open ocean and moved into inshore waters, but by the time of the late Miocene–early Pliocene cooling event even these sharks were in trouble.

The problem with such large sharks is that they need a lot of calories just to stay alive. The Zurich team worked out the food requirements of their 18-metre-long megalodon; so, if you want to know what makes a megalodon tick, here goes. Basing their calculations on the extant white shark, they found that their long megalodon would have required twenty times the daily food intake of a modern white shark. In numbers, this means 98,000 kilocalories per day for a megalodon, compared to 5,000 kilocalories for a modest-sized white shark weighing about 900 kilograms (2,000 lbs). To meet this energy demand, the megalodon would have had to consume about 21 kilograms (46 pounds) of blubber each day, assuming that 30 kilograms (66 pounds) contains 200,000 kcal. This also assumes 70 per cent of food intake is actually assimilated. It means the megalodon could eat about 80 per cent of a long-snouted dolphin weighing, say, 120 kilograms (260 pounds), but as little as 0.1 per cent of an orca-sized whale, to satisfy its daily needs. It might also snack on just 1.4 per cent of the liver of a 7-metre (23-foot)-long white shark. These figures may not sound much, but bear in mind the shark would have to eat this amount of food every day and the food might not be available every day. However, if the giant shark could eat its largest prey, for example a large baleen whale, in its entirety then it could go about its everyday life for nearly two months without having to feed again.

These sharks also homed in on the most nutritious part of the whale's body. Sperm whales and their relatives, for example, have a bulbous forehead containing huge amounts of fat – the spermaceti organ. This is thought to be an acoustic lens that focuses high-frequency sounds that the whale uses to interrogate its environment. It appears to have been irresistible to megatooth sharks, and research has shown that they did, indeed, take advantage. Looking at the bite marks on several ancient sperm whale skulls, a small team of researchers from Switzerland, USA, Italy and Peru found that megatooth sharks bit out huge chunks of the fat-rich forehead, but bite marks also showed that smaller sharks came along afterwards and nibbled at anything that remained. Surprisingly, some of the whales survived. The sharks only stripped way the fat close to the nasal organs and nothing else. Towards the end of the megalodon's reign, though, it became clear that this sort of food was beginning to become scarce. Small whale stocks were dwindling and were shrinking into smaller, more fragmented populations. And, having developed an elevated body temperature, the megalodon would have needed high-energy foods, like the blubber from baleen whales, to be able to generate and retain warmth.

The megalodon lived for close to 19 million years, its demise thought to have occurred around 3.6 million years ago, in the middle of the Pliocene epoch, when the continents and oceans were close to their present positions and many marine creatures were already modern species. It was, curiously, the warmest phase during ongoing global cooling. Even so, the changes in climate must have had an influence, especially a general drop in global sea surface temperatures, but scientists believe there was also another factor at work. None other than the white shark was in the ascendency.

For the best part of 150 years it was believed that the white shark was descended from the megalodon. The evidence was pretty flimsy: both species have triangular teeth with serrations, and that was about it. As a consequence, most depictions of megalodons have been of grossly puffed-up white sharks. However, recent research suggests an alternative theory, and that the two species are not so closely related. White sharks, it suggests, are allied more to ancient mako sharks.

One of its ancestors is thought to have been the broad-toothed mako shark *Carcharodon hastalis*, a shark that maxes out at about 8 metres (26 feet) long. While primarily a fish eater, fossil evidence from Italy suggests it might have chomped on dolphins from time to time. The dolphin in question was *Astadelphis gastaldii*, which lived from the Pliocene but became extinct in the Quaternary epoch. This particular specimen was 2.8 metres (9 feet) long and in the region of 3.8 million years old, and it had been long forgotten in a museum in Turin. However, when researchers from Pisa dug it out and dusted it off, they found a series of crescent-shaped bite marks that could have only come from a shark. At first, they thought it was the work of a white shark, but closer examination revealed that the bite marks were made by teeth that were not serrated. Two candidates were the extinct broad-toothed mako and its living cousin the shortfin mako, the size of the shark, about 4 metres (13 feet) long, making the former a more likely candidate than the latter. What the research was able to show, however, was precisely how the attack progressed and therefore how the dolphin died.

The toothmarks on the dolphin's jaws, ribs and vertebrae indicate the series of events. The shark approached from behind and to the right of the dolphin and grabbed the abdomen on its right side. The dolphin struggled and in doing so caused even more damage to its body as the shark shook its head from side to side and its teeth sliced away a fair chunk of flesh. The dolphin must have lost blood rapidly and died or was in shock. It began to roll onto its back, whereupon the shark grabbed it again just behind the dorsal fin, leaving a second series of bite marks in the dolphin's vertebrae. The shark then fed on the soft tissues, but it wasn't alone. Smaller bite marks indicate that, once the shark had left, small scavengers came to feed on the body. It was a remarkable bit of palaeontological detective work.

The broad-toothed mako shark, though, features in another bit of sleuthing, which revealed more of the evolution of the white shark. The evolutionary line can be followed in a series of steps that can be traced in the fossil record even though the evidence is mainly teeth, scales and vertebrae as the rest of the shark is rarely preserved. However, researchers at the University of Florida and other associated universities got lucky. They unearthed several fossilised vertebrae and a jaw with teeth intact from an extinct shark species new to science.

It lived during the Late Miocene, 6–8 million years ago, and was given the scientific name *Carcharodon hubbelli*, also known as Hubbell's white shark. The scientists proposed that it was a halfway house species between the now extinct broad-toothed mako *Carcharodon hastalis* and the living white shark *Carcharodon carcharias*.

Again, teeth played a key role. The extinct mako had sharp, triangular teeth similar to the white shark, and, at up to 7.5 centimetres (3 inches) tall, they are the second largest after megalodon teeth, but there are no serrations, indicating a predator that mainly eats fish (and the occasional dolphin). The white shark has serrated teeth, all the better for carving up marine mammals. Hubbell's white shark, though, had teeth which were serrated but not as sharp as either the mako or the white shark's teeth. From evidence like this, the researchers suggested that the extinct broad-toothed mako was the ancestor of the white shark and not the megalodon, with the extinct Hubbell's white shark as a transition species, which makes the white shark a meat-eating variety of a fish-eating mako shark and not a miniature version of the megalodon.

Whatever theory proves to be correct, the white shark turned out to have a thing going with its great-uncle. The megalodon might well have been indestructible during its heyday, but later, when things got tough, it is thought to have been outcompeted by the smaller and nimbler white shark. Evidence comes in the form of isotopes in teeth again, but this time an isotope of zinc – zinc-66. Plant eaters tend to have more zinc-66 while meat eaters have less and, the higher up the food chain an animal sits, the lower the values of zinc-66 in its body, the lowest levels shown by the apex predators right at the top.

Checking the zinc levels in tooth enamel and the dentine layer below, of both the megalodon and the white shark, especially where their ranges overlapped, revealed that they probably ate the same foods – marine mammals, such as small to medium-sized whales and seals. So, when food became scarce due to a big drop in populations of marine mammals, competition with the smaller white shark could have been one reason for the megalodon bowing out. It would have needed many more calories than its smaller relative, so, while the white shark retained its position at the top of the food chain, the megalodon was relegated and eventually became extinct, but this probably was not the only cause. Changes in ocean currents, for instance, could have had a bearing too, such as when the Central American Freeway between North and South America was blocked by the formation of the Isthmus of Panama, upsetting the circulation in both the Pacific and Atlantic Oceans. And then, of course, there's our old friend climate change, which underlies many of these things.

In another similar study by researchers from Germany's Max Planck Institute and several US universities, tracking zinc isotope levels revealed how a species like the megalodon would have declined due to changes in food availability. The focus was on several *Otodus* species that lived in the Atlantic Ocean during the Miocene, and it showed that zinc isotope ratios in their teeth changed over time. One of them was *Otodus chubutensis*, a megatooth species that was thought to be the immediate ancestor of the megalodon. About 16 million years ago, it was an undisputed apex predator. This was manifested as low levels of zinc-66 in its teeth. However, by 5 million years ago, the level of zinc-66 had risen significantly, indicating the shark had dropped back in the food chain and, basically, the species, like its descendant, was on its way out. Ironically, with the giant sharks and small whales out of the picture, meaning the competition for zooplankton and small fish was removed, the larger whales, particularly the baleen whales, were able to grow to tremendous sizes; witness the blue whale, the largest animal ever to have lived on earth.

The white shark, however, survived all these climatic changes and the extinction of the small baleen whales. It clearly was more flexible in what and how it hunted than the megatooth sharks and could adapt to changing circumstances. The modern white shark probably had its origins in the south-east Pacific Ocean, where scientists have discovered what they believe to be the first known palaeo-nursery site. While the teeth from subadults and adults dominate at two fossil white shark sites nearby, namely Pisco in Peru and Bahia Inglesa in Chile, at the fossil site at Chile's Coquimbo there are mostly juvenile teeth. During the Pliocene, between 5 and 2.6 million years ago, it must have been a white shark nursery. Along with the large finds of adult and subadult teeth in the region, it appears that the Chile–Peru coast had a much larger population of white sharks than it has today; in fact, nowadays, they're quite rare here.

While fossil teeth of early white sharks can be found in rocks laid down in the mid-Miocene, the first fully developed serrated teeth, which is a characteristic of modern white sharks, appeared at the Miocene–Pliocene boundary, which means the species has many millions of years to go to equal the success of the megalodon.

However, there are people who think that the megalodon

is still living and hiding away in the deep ocean. They can't, though, explain the lack of fossil teeth from about 3.6 million years ago until today. There is some questionable fossil evidence that the megalodon could have survived until about 2.5 million years ago, but generally scientists currently believe the older exit date to be the correct one. The megalodon is dead, but the white shark lives on… at least for now.

PART III

A THOROUGHLY MODERN SHARK

CHAPTER 6

TRANSOCEANIC JOURNEYS

Nicole was an extraordinary shark. Named after the Australian actor Nicole Kidman, a known friend of sharks, this 3.8-metre (13-foot)-long female white shark became a record breaker. On 7 November 2003, she was fitted with a PAT (pop-up archival transmitting) tag in Haaibaai, near Dyer Island, Gansbaai, South Africa, by researchers from the White Shark Trust and the Marine and Coastal Management Department of South Africa, and then nothing was heard from her for over three months. On 28 February 2004, however, the link attaching her tag to her dorsal fin broke, and the tag popped up and began transmitting information to a satellite orbiting overhead.

When the team came to analyse the data, such as location, depth and seawater temperature, it found that Nicole had first headed towards the south-west, probably caught up in an Agulhas ring – one of the great vortices that form in the fast-flowing Agulhas current, which bathes the coast of south-east Africa. The rings spill into the Atlantic Ocean, but Nicole must have used the anticlockwise swirl to sweep around and to head eastwards into the ocean basin. The big surprise was that by the end of this leg of her journey, she had reached a point just offshore of Western Australia, 37 kilometres (22 miles) to the south of Exmouth Bay. In just ninety-nine days, she had travelled 11,100 kilometres (6,900 miles) across the entire breadth of the Indian Ocean.

With her tag released, she could not gather and transmit any more data; but that was not the end of the story. Again, she disappeared off the radar, only to reappear back in South Africa on 20 August 2004. This time she was spotted as part of a photo-recognition programme set up by Swiss-born researcher Michael Scholl of the White Shark Trust. A telltale pattern of nicks on the trailing edge of her large dorsal fin gave her away. She had made the round trip of at least 20,000 kilometres (12,500 miles) in less than nine months, a world record for the fastest return migration of any known marine animal. What was even more remarkable was her speed. The researchers calculated that she must have had a minimum speed of 5 km/h (3 mph), comparable to the cruising speed of the fastest tunas… and there's more. During the journey, Nicole made a record-breaking dive to 982 metres (3,222 feet), the greatest depth for a white shark at the time, and experienced an unexpectedly low sea temperature of just 3.8°C (38.8°F).

The data also showed that, for close to 60 per cent of her journey, Nicole swam within 5 metres (16 feet) of the surface, prompting shark biologists to speculate that she was using celestial cues, such as the moon and the sun, as well as the earth's geomagnetic field, to find her way. Her track was remarkably straight, so she certainly seemed to know where she was going, and she was getting there in the fastest way possible.

One of the big unknowns, though, is why Nicole travelled to Australia and back in the first place. She had always appeared at Gansbaai at some time between June and December and was familiar to the researchers there; but she disappeared for the rest of the year, which prompted some to speculate that she has been making the same journey every year. Before the age of tagging and tracking we would never have known. But why? Some suggest she goes to find the shark of her dreams, but a 3.8-metre female white shark is not ready to mate, not until she's closer to 4.5–5 metres (15–16 feet) long. Could she be making dummy runs, so she knows the ropes, ready for the time she comes of age? There are breeding populations of white sharks along the coasts of Western and South Australia, so that could be an attraction.

Other scientists have suggested her journey is food orientated, but what food? There are a few sea lion colonies with increasing numbers in the area of the Western Australia coast where Nicole pitched up, but they're not a big draw, and she would be there at the wrong time of year for scavenging on humpback whale carcasses, so it must remain an unanswered question for the moment, all part of the intrigue surrounding this species which has prompted a flurry of new and revealing research.

One of the things that stands out about Nicole's journey is the speed with which she covered such an impressive distance, as mentioned above. This is all down to sharks having adaptations that enable them to move almost effortlessly through the water. For one thing, white sharks have a fusiform-shaped body that is rounded and tapering at both ends. It's hydrodynamically an ideal shape, reducing drag and lowering energy requirements to make the shark move. The shape of the white shark's tail or caudal fin is also relevant: it's scimitar shaped, with the top part a touch longer than the bottom part. It's the sign of a fast shark, the shortfin mako being the fastest with top and bottom of almost equal length. It can power along with a burst speed of about 70 km/h (45 mph). The white shark, though, is up there with the fastest sharks, with a burst speed of close to 55 km/h (35 mph). By contrast, slow-swimming sharks have the top part of the caudal fin significantly longer than the bottom part, which helps them keep up their snout when swimming slowly.

The structure of the fin and the part that joins it to the rest of the body – the peduncle – are also significant. The two together form a dynamic locomotory structure. The caudal peduncle is rigid, the stiffness facilitating the movements of the caudal fin. This is achieved internally with a thick layer of adipose tissue, which stores energy as fat, reinforced by cross-woven collagen fibres, which provide structural support and mechanical strength. The caudal fin itself is in two parts, with the upper and lower lobe structured internally in a slightly different way. Muscle bundles separated by connective tissues, together with a flexible rod-like notochord, make up the bulk of the upper lobe. The leading edge is reinforced with adipose tissue, so the cross-section of the upper lobe is an advanced hydrofoil. The lower lobe consists of fin rays separated by connective tissue and has greater bending stiffness than the upper lobe, although the upper lobe has greater control of movement. The angles of the fibres along the span of the upper lobe also resist bending stresses that the lobe is subject to when the tail is moved, and they store strain energy which is probably useful in powering the recovery stroke. This complex fibre architecture in the caudal fin and caudal peduncle gives rise to a potential elastic mechanism in the way the white shark swims and so is probably making a contribution to energy conservation. There is also a single keel on each side of the peduncle that assists in the hydrodynamics of the shark.

The workings of the caudal fin, however, appear to be more important to young white sharks than mature ones. The younger sharks rely on hydrodynamic lift, care of the wing-like pectoral fins, the necessary speed generated by the un-usually large span of the tail to stay 'up' in the water column. It also enables them to pursue fish at intermediate speeds with the minimum of effort. Mature individuals rely more on buoyancy, care of the lipids in their large liver, to prevent them from sinking, especially when swimming slowly and feeding or scavenging on mammalian prey.

The white shark's large dorsal fin is also special. It is filled with dermal fibre bundles that extend into it from the body. They are staggered, with fibres lying at angles of more than 60 degrees, rather than in straight lines or rows, giving tensile strength without impeding fibre movement. Compared with the dorsal fins of other large species, such as tiger sharks, the fibre bundles are more densely packed, and they extend the biggest distance into the fin. Scientists from the University of KwaZulu-Natal believe this arrangement indicates that the white shark's dorsal fin is a dynamic stabiliser, responding to the speed the shark is swimming at. When swimming fast, the problems of yaw and roll are greatest. The hydrostatic pressure within the shark increases and the fibres around the entire body, including those in the dorsal fin, become taut, there-by stiffening the fin. When swimming slowly and also while feeding, the hydrostatic pressure reduces, the fibres slacken and the muscles are able to exert greater bending force and therefore give greater flexibility. The trade-off is greater stiffness for fast swimming against that flexibility, but it means the white shark can swim unbelievably fast when it needs to.

Another adaptation is the skin. If you brush your hand over a shark's skin from head to tail, the skin feels relatively smooth but move your hand the other way and it's like sandpaper. This is because the skin of most sharks is covered by dermal denticles, otherwise known as skin teeth. Structurally, they are just like tiny teeth, between 0.2 and 0.5 millimetres (0.01–0.02 inches) across, with enameloid and dentine outer layers and a central pulp cavity and they are closely packed together in longitudinal ridges aligned along the axis of the body; but they are more sophisticated than mere teeth. Every shark species has a different-shaped dermal denticle, depending on its lifestyle. The white shark, for example, has particularly small, closely packed dermal denticles so it doesn't feel as rough as other shark species. However, the basic structure of these skin teeth is the same – a flattened crown and an elongated neck that extends to an expanded base which is embedded in the dermis.

Under the microscope, white sharks appear to have flat, overlapping, V-shaped denticles, another mark of a fast swimmer. Smooth skin causes drag, which would slow down the shark, so it would have to use up excess energy to move through the water. The shark's rough skin, however, guides the water over its body in the most effective way, controlling the flow close to the skin, and so reducing drag. The denticles also create low-pressure zones, known as leading-edge vortices, which suck the shark forward, resulting in more forward thrust. These adaptations enable the shark to swim faster and more efficiently. Swimwear manufacturers have copied the concept with claims that it gives the wearer that extra edge in competitions, although recent research has shown that the fake denticles themselves don't work. Real shark skin is more dynamic. Unlike swimwear denticles, a shark's dermal denticles can be moved: they can bristle to an angle of between 30 and 50 degrees, depending on the species. They follow the bending and flexing of the shark's body as it swims, and thereby shift the direction of the flow of water close to the skin to gain the maximum effect.

The shape of the dermal denticles might also be different depending on where they are located on the shark's body. In the thresher shark, for example, denticles on the leading edge of the fins and tail tend to be flattened and more rounded, with few ridges, whereas those on the trailing edge are triangular with many ridges. Maybe white sharks have the same arrangement, but the research has yet to be done.

White sharks, however, have another locomotory trick up their sleeve. They keep their swim muscles warm, with an amazing adaptation known as the *rete mirabile*, the Latin for 'wonderful web'; in fact, they have at least two, one on each side of the body. This heat exchange system consists of a honeycomb of arteries and veins in close proximity, allowing warm, deoxygenated venous blood on its way from the muscles to the gills to transfer heat to the cooler oxygenated blood moving in arteries from the gills to the rest of the body. In this way, the shark retains heat generated by its own muscles, keeping its body temperature up to 15°C (27°F) above that of the surrounding seawater. Organs that benefit include: the eyes and brain, so the white shark is more alert than many of its rivals; the stomach, so it's able to digest a meal rapidly and be ready to hunt again quickly; and the swim muscles, so it's able to retain its speed even when the seawater temperature has dropped to 3°C (37°F). Technically, the white shark is a mesotherm or endothermic poikilotherm, which means its body temperature is not constant, like that of humans, but it can be regulated. If it needs to save on energy, for example, it can actually lower its body temperature as well as raise it.

Energy conservation, of course, is vital for an animal that embarks on such long migrations. White sharks have ways in which they conserve fuel. Firstly, they stock up on fats obtained from seals and sea lions. These are stored in their massive liver, which can account for up to a quarter of the shark's body mass. In an experiment with four sharks tagged off California, researchers from Stanford University and Monterey Bay Aquarium found that the key to their long-distance journeys was 'drift diving'. This is when the sharks descend passively, their momentum carrying them forwards, like an underwater glider. It's a behaviour which many pelagic sharks, such as the blue shark and the oceanic whitetip shark, adopt to make the best of their fuel reserves. The researchers were able to measure the rate at which the white sharks sank during a drift dive and so estimate how the amount of oil in their livers changed as their journey progressed. More oil in the liver made the sharks more buoyant, as oil is less dense than water, and so they made a slow descent. Less oil meant they were less buoyant and made a fast descent. During the sharks' journeys, especially those of more than 4,000 kilometres (2,500 miles), they discovered that buoyancy decreased as the journey progressed, indicating a gradual depletion of their fuel reserves.

They were running out of the fuel that they obtained before setting off, which means they must feed well before embarking on a long journey, as they don't snack on the way. The shark's cartilaginous skeleton is also much lighter and more flexible than, say, bone, reducing body mass, which also helps keep the animal afloat.

Drift diving is supported by dynamic lift. The white shark has quite large pectoral fins which are configured like an aircraft's wing, providing lift that helps keep the shark up in the water column. It and pelagic sharks like it acquired their large pectoral fins about 93 million years ago in response to an increase in carbon dioxide levels that caused the oceans to heat up and oxygen levels in the deep sea to drop. The mainly bottom-dwelling sharks were forced to move closer to the surface and become pelagic species. This expansion into the open ocean probably involved extended hours of sustained fast swimming to catch food or stay out of trouble, which led to a selection pressure in favour of efficient movement via higher-aspect-ratio pectoral fins, a sign that adaptive evolution has been taking place.

It was these kinds of adaptations that enabled Nicole to complete her return trip between South Africa and Western Australia in record time. Until shortly before this event, white sharks were thought of as a coastal species; after all, that's where people mostly saw them, but a couple of years before Nicole's revelations, scientists on the west coast of North America discovered that the white sharks they were seeing close to the Farallon Islands, Año Nuevo Island and Guadalupe Island also had long-distance aspirations. While the sharks focused on seal and sea lion rookeries on these coastal islands at times of the year when youngsters were about, they spent the rest of the time offshore, and a considerable distance offshore at that.

Before this, scientists were convinced that white sharks of both sexes and all ages never ventured far from the place of their birth, but now we know that, while young of the year and juveniles prefer the coast and don't move far from their birthplace, this is not the case for older sharks; in fact, quite the opposite. In the eastern North Pacific, they head out to sea, following a migration corridor that takes them to a special patch of ocean about halfway between Baja California and Hawaii. Scientists without a sense of humour named it the Shared Offshore Foraging Area (SOFA), while those with called it the White Shark Café. Whether it really is a foraging and feeding area is far from clear, despite attempts to find out, but the sharks are generally on extended visits, the length of time depending on sex and age.

They leave the mainland coast in spring and summer, travelling generally westwards (depending on from where they set out) for about twenty-five days. On arrival at the Café, males and females behave in very different ways. Male sharks, most more than 3 metres (10 feet) long, make for the core of the area, where they oscillate between the surface and the depths in a rapid and random manner about 120 times a day. The females range over a larger area, and they follow a typical diel vertical migration pattern, i.e. close to the surface at night, where sea surface temperatures are between 20 and 26°C (68–79°F), and down to depths of 300–500 metres (1,000–1,600 feet) and sometimes even 600 metres (2,000 feet) by day, where the temperature is close to 4.8°C (40.6°F). The females rarely enter the core area, spending about 4 per cent of their time there. What with the males oscillating and the females dipping in occasionally, it feels like the male sharks are performing in a lek. On land, this is usually an area – the arena – where males of a species gather to display to visiting females, like prairie chickens on the North American prairies. However, there is no evidence so far to suggest that the Café is a courtship and mating area.

Food is a possible draw. It was once thought that the open ocean was something of a desert, but now we know otherwise. Although the surface waters are what's known as an 'epipelagic cold spot', with little phytoplankton and low productivity compared with other parts of the eastern North Pacific, a layer of phytoplankton has been found at a greater depth than satellites can penetrate, so the base of a food chain is present, albeit deeper down, and there are a whole range of creatures that take advantage of this unexpected oasis. Aside from the migrants that are passing through, such as tuna and various billfish, several species of pelagic squid come here to spawn, and white sharks are known to catch squid. Also, a bony fish, the pomfret, one of the butterfish family, spawns in the Café. It feeds in the sub-Arctic waters in summer and migrates to the subtropical North Pacific Gyre in winter to spawn. So, there's a fair number of potential prey species present. The only thing is that these Café creatures are not as energy-rich as the seals and sea lions at the coast, and scientists have noticed that some sharks return from the Café significantly thinner than when they went out. Nevertheless, even though white sharks can fast for long periods, they must be eating something out there or they wouldn't survive.

Males and immature females might remain in the Café for

up to five months before returning to the mainland coast, but mature females, more than 4.5 metres (15 feet) long, stay for much longer – up to sixteen months. It's thought they are pregnant and spend the best part of their eighteen-month-long confinement in the area of the Café, out of the way of amorous males. When the time comes to give birth, they head back towards where they came from. Guadalupe Island females, for example, make for the island, but they don't stop off there. They carry on to the mainland and sheltered bays on both sides of the Baja California peninsula, such as Sebastián Vizcaíno Bay on the central Pacific coast of Baja. Here, they are thought to give birth, as young-of-the-year white sharks are often observed or caught in the vicinity of these sites.

Not all California's mature females stop off at the Café. Some keep going all the way to Hawaii, a straight line distance from Baja of more than 4,000 kilometres (2,500 miles). The attraction here is mainly whale carcasses, from humpback whales that migrate here from Alaska and other points along the Pacific Northwest to drop their calves or mate. Sperm whales come too, in search of squid, and two of the world's largest living white sharks were seen here feeding on a sperm whale carcass. Spinner dolphins and monk seals also appear on the menu, and there is another SOFA to the south of the island chain.

Some of California's older sharks, having forsaken a trip to the Café, travel to the north, some reaching south-east Alaska and the Aleutian Islands, where the sea surface temperature can be close to 5°C (41°F) in winter and a more balmy 16°C (61°F) in summer. These characters tend to be the larger sharks in the size range 3.8–5.4 metres (12.5–17.7 feet), well capable of functioning in the more modest temperatures. One individual was observed near Cordova, not far from the mouth of Alaska's Copper River, making it one of the most northerly known white shark sightings. Food, according to the stomach contents of captured sharks or those washed ashore, includes salmon and Pacific halibut, along with Steller sea lions and harbour seals, which means these sharks are competing with orcas and salmon sharks (a close relative of the white shark) for some of what's available.

On North America's east coast, the main white shark migration is generally north–south for all sharks. In summer, they might reach Nova Scotia and Newfoundland, passing Cape Cod on the way northwards, where they stop off for a seal snack. Cape Cod is thought to be the biggest concentration of white sharks in the world. In winter, most head south, with the coast of the Carolinas a favourite place to overwinter, although individuals go on to Florida and even turn right to enter the Gulf of Mexico, some reaching Mexico's Gulf coast.

The OCEARCH project has revealed that some white sharks travel eastwards across the Atlantic. A large female, which the researchers named Lydia, travelled all the way to the Mid-Atlantic Ridge. The point at which she crossed was closer to Europe than North America, but eventually she turned back towards the USA. British readers and white shark aficionados were disappointed she didn't make it all the way, but could white sharks be going the extra mile beyond the Mid-Atlantic Ridge and on to the British Isles? There's a recent sighting that suggests they could be.

In July 2024, two fishermen plying Ireland's Galway coast reported what they thought was a white shark circling their boat. The fish was about 3 metres (10 feet) long, and video they took on their smartphones indicates something which could well be a white shark, but the sea's surface was too choppy and the water too cloudy to firmly identify it as such. The behaviour of inspecting a boat and its crew, though, certainly smacks of white shark.

It was the latest in a whole raft of supposed sightings of white sharks by holidaymakers and sports fishermen that appear without fail in British newspapers every summer, although there has still to be even one confirmed report, despite the UK's Shark Trust logging more than 100 sightings. Nine of these have been described by Richard Peirce, a trustee of the Trust, as 'credible'. They range from Cornwall to Scotland, and maybe it is worth taking a minute to look at what people have seen and what bearing these sightings might have on our knowledge of the distribution and movement of white sharks in the north-east Atlantic.

Although there have been reports since the nineteenth century, the earliest credible sighting was in 1970, when a shark angler based in Looe, Cornwall, was pulling in his chum bags and fishing lines. A large shark poked its head out of the water, checked out the fisherman and slipped below the waves. He thought, but could not be sure, that he had seen a white shark.

Then, three interesting sightings were made in August 1999. First, the sports fishing boat *Blue Fox* was off Cambeak Head, near Crackington Haven, on Cornwall's north coast. One of the anglers had caught a tope and was reeling it in when a large shark, estimated at about 15 feet (4.6 metres) long, appeared. It rolled slightly, exposing its white belly, before disappearing back into the depths. There were cameras on board but, during the brief encounter, there was no time to use them. However, there was more. The next day, anglers were on the *Blissful*, fishing at the same spot, when a large shark, estimated to be 17 feet (5.2 metres) long, came up and bit off two-thirds of another tope being reeled in. The fishermen had seen porbeagles, makos and basking sharks but they thought their shark was not one of those, and the skipper of the *Blue Fox* had spent time in South Africa and was familiar with white sharks. He thought the shark that had visited both boats was a great white shark. And there the matter rested until a couple of weeks later, a lobster fisherman off Tintagel Head, about 18 kilometres (11 miles) to the south-west, reported a large shark entangled in his lobster pot lines. It had a slate-grey back, a white belly and a crescent-shaped mouth with triangular teeth – a reasonable description of a white shark. However, as the carcass was deemed to have no commercial value, it was cut loose; again, no photograph.

In 2002, another lobster fisherman was off The Quies, two small islets off Trevose Head to the south-west of Tintagel. He spotted something breach and went over to investigate. There was lots of blood and what he described as 'seal bits' in the water. Now, killer whales breach when attacking seals, and they frequent the west coasts of the British Isles. White sharks have been seen to breach spectacularly in South Africa, but less frequently elsewhere. The fisherman had seen killer whales in the area before and felt that the animal making the attack was not one of them. Two days later a yachtsman spotted a large shark that he thought was not a basking shark. It followed his boat for much of his journey between Padstow and Newquay, which meant he passed through the waters where the breach had occurred.

While Devon and Cornwall are well known as shark hot-spots, there is also a fair bit of activity around the coasts of Scotland, and the next credible sighting was at the western edge of the Summer Isles, not far from Ullapool on Scotland's west coast. It was a sunny, hot day at the beginning of July 2003, and two divers had just left the water near Black Rock and two others were about to dive, when a large shark came to take a look at them. At first, they saw the dorsal fin some distance behind the boat – a RIB – and thought it must be a basking shark. The two divers who were kitted up thought it would be fun to swim with it. They switched on the engine to motor over to its position, but the shark suddenly changed direction, almost aggressively, and made a beeline for the boat, a common behaviour of white sharks and not basking sharks. It swam alongside for about fifteen seconds, and it felt as if it was giving the occupants the once-over, again a white shark trait, and then dived and swam away. One of the divers was a marine biologist, who gave a detailed description of the shark to BBC Earth. He estimated it to be more than 5 metres (16 feet) long. It had a large, pointed snout with small eyes, white belly and grey back, a large, broad-based triangular anterior dorsal fin, and no sign of the unusually large gill slits of the basker. He was convinced he had encountered a white shark. White sharks, however, don't have small eyes, the same with the other mackerel sharks – porbeagles and makos – that occur around the British Isles.

Two weeks later, two fishermen were tending their nets off

north-east Scotland. One of them, George Carter, submitted a short paper to the *Glasgow Naturalist*. It describes how he was fishing for codling off the coast of east Caithness, close to his home port of Lybster, when he spotted two large fins, some distance apart, in his net. His first thought was that they were two dolphins, but when he approached more closely, he could see he had trapped a large shark. It was well entangled and began pulling the net and the boat for about 200 metres (220 yards). Carter's next thought was that he had caught a basking shark, but he had seen many of those and this shark was different. He could also reject a thresher shark as he could see the tail, and a porbeagle because it was just too big, about 5.5 metres (18 feet) long, roughly the same length as his boat. After more struggling, the net and shark sank to the seabed, so Carter towed it into shallow water close to a cliff face and attached an anchor and buoy. He then went to get help to recover his net. When he returned, the shark was at the surface once again. Carter managed to take one picture with his new digital camera, at which point there was 'an explosion of movement and the shark very rapidly shot away from the boat, free of the net'. Carter and his colleague were both very sure the creature was not a *muldoan*, the local name for the basking shark. Carter described it as 'broad-headed and the steely-grey dorsal surface was smooth … and the eye was dark', but he could not check the colour of the underside as the shark was swathed in net and it was too big to move. On returning home, he checked several books and guides and concluded that what he had caught was a white shark, but he could not be 100 per cent sure… however, he had a photograph and a sharp one at that. The anterior dorsal fin belongs clearly to a large shark and probably a lamnid – porbeagle, mako or white shark. There is no obvious white patch on the bottom of the trailing edge, which, perhaps, eliminates the porbeagle, but without more features visible it is very difficult to pin down a species.

Next up is a large shark, estimated to be 5.5–5.8 metres (18–19 feet) long, which was caught in a fish trap near the Pentland Firth between Orkney and north-east Scotland in December 2003. It snapped at fish in the trap, was photographed, and then escaped. In July 2005, anglers were trolling for pollock off North Uist, in the Outer Hebrides, when a large shark fitting the description of a white shark surfaced briefly close to the boat, and the following year another large shark was caught in a trawl net, but the photograph taken was not good enough for a positive ID. A few years later, in 2014, a fisherman was off Falmouth on Cornwall's south coast and spotted a large shark, which came alongside his boat. He took a video as it passed the boat, just under the surface, but it was taken into the sun, so the light reflected off the water and so the ID again is inconclusive. It could have been any one of the lamnid sharks, and it just goes to show how difficult it is to identify sharks in their natural setting around the coasts of the British Isles.

One thing common to most of these sightings, however, is that in almost every case a boat's occupants are convinced they have seen a white shark, and many of them are well versed in fish identification. There's certainly plenty for white sharks to eat in the waters of the British Isles – an estimated 50,000 harbour or common seals and 120,000 grey seals, as well as all the smaller sharks, such as tope and smoothhounds, along with mackerel and nowadays even bluefin tuna, which would be a strong attraction for white sharks. Also, conditions are favourable. The waters in this part of north-west Europe have the same kind of surface temperatures as those off South Africa, a known white shark hotspot. However, the presence of white sharks around the British Isles has still to be confirmed. Despite all the press hype each summer, when virtually any species of shark sighted by holidaymakers off popular beaches is claimed to be a white shark, the reality is that it's probably not.

However, if – and it's a big 'if' – a white shark is shown to have entered UK waters, where could it have come from? There are two options: it could be from the north-west Atlantic, where the most northerly reports are from the Gulf of St Lawrence, near Newfoundland, on Canada's east coast, which is on the same latitude as the English Channel, or from the north-east Atlantic stock, of which there is very little migration data.

However, two of the most northerly records in this north-east part of the ocean are from the northern Bay of Biscay. One is from close to the mouth of the Loire, and is known from a set of preserved jaws, while the other report is from May 1977 of a 2.1-metre (7-foot)-long juvenile female great white caught in the Pertuis d'Antioche, a strait between two islands – Ile de Ré and Ile d'Oléron – and the French mainland, close to La Rochelle. This is just 310 kilometres (193 miles) from Land's End as the crow flies (or the shark swims). White sharks have also been recorded on the Atlantic coasts of Spain and Portugal – for example, one was spotted swimming off La Coruña, Galicia, north-west Spain, in September 2021 – but most sightings are along Mediterranean Sea coasts, mostly in the west, which are frequented by white sharks of all sizes, including some notable big ones. They have been recorded along all coasts in the western basin of the Mediterranean, such as in the Gulf of Lyon, and they have also appeared off Spain's Balearic Islands and in the Catalonian Sea in the vicinity of Valencia.

On the other side of the western basin, white sharks occur in the Tyrrhenian Sea off the west coast of Italy, including the Gulf of Genoa. In this western basin, phytoplankton biomass and primary production is relatively high compared to further east so there should be good foraging for sharks. In the central basin, the Sicilian Channel and the Adriatic and Ionian seas are areas of high productivity, while the Adriatic and Aegean seas and the Sicilian Channel, especially the coasts of Tunisia and Libya, are all thought to be birthing and nursery areas. In recent years, pregnant females have been caught at these locations. In 1992, for instance, a 5-metre (16-foot)-long female with two embryos was intercepted at Tunisia's Cape Bon, a promontory that projects into the Sicilian Channel, and a 5.87-metre (19.3-foot) female carrying four embryos was caught a little to the south in the Gulf of Gabès in 2004. Both were by-catch from commercial fisheries.

Unfortunately, little tagging research has been carried out in the Mediterranean so migratory movement must be inferred from distribution data, obtained mainly from the media, fishermen and the general public. There is, however, one thing that stands out: many white sharks, especially the larger individuals, are found mainly in areas where Atlantic bluefin tuna are migrating to their spawning grounds. Whether they wait and ambush them or follow them across the Mediterranean is not entirely clear.

The tuna migrate huge distances. They leave their feeding

sites, close to the coasts of Greenland, northern Norway, the maritime provinces of eastern Canada and also west Africa, and head to the Mediterranean, where they spawn. After they pass through the Strait of Gibraltar, avoiding the long-line fisheries and orcas that patrol the channel, the majority go to the Med Gate, an area of sea in the southern sector of the western basin stretching from Gibraltar to the Algeria–Tunisia border. A large white shark was caught here, between Al Hoceima and Nador off the north coast of Morocco, in 2017, and another was videoed in 2013. Some tuna remain in this area, but the vast majority go to the Balearics and the Gulf of Lyon. Here, a combination of the North Mediterranean Current, the North Balearic Front and discharge from the River Rhone creates an area of high productivity which attracts large marine predators. It also entices tuna to stay all year round, for not all return to the Atlantic. It means there are two populations of tuna – residents and migrants – and so white sharks should have a food source throughout the year, although it's thought that in recent years dolphins have become an important component of their diet. Bottlenose dolphins follow trawlers to snap up the fish that escape from the nets, especially in the Adriatic. White sharks must be joining the procession, and evidence for them hunting live dolphins, rather than simply scavenging, comes from a bottlenose dolphin with two fresh bite marks that was seen near Lampedusa in May 2006. From the size of the bite marks, the shark responsible was thought to have been more than 4 metres (13 feet) long.

All these locations, unsurprisingly, are hotspots for the Mediterranean's white sharks. Some can be pretty hefty, and they make their presence felt from time to time. In the

Adriatic in 1998, a sports angler and his son were fishing close to an oil rig off Rimini when they had a bite. Reeling in the catch, a medium-sized thresher shark, and tying it alongside the boat, they looked up and suddenly spotted a black dorsal fin heading straight for them. '*Uno squalo bianco!*' shouted the father, just as it slammed into their hard-won catch. Thrashing its head from side to side and slicing into the smaller shark with its triangular, knife-like upper teeth, it tore away huge mouthfuls before disappearing below just as quickly and silently as it arrived. The fisherman filmed the event, possibly the first footage of a white shark in the Mediterranean, and, if you care to, you can watch it on YouTube. Researchers who have viewed the video have suggested that the shark was about 6 metres (20 feet) long, close to the maximum size for this species. It was a day the two anglers will never forget.

Another Mediterranean region that has attracted scientific interest in recent years is the Turkish Aegean coast. White sharks disappeared from here several years ago, but now they're coming back and they're dropping their pups here. Youngsters with a length of between 80 centimetres (32 inches) and 1.6 metres (5 feet) have been caught in the Gulf of Edremit, and, if the local fishery is anything to go by, the bay has plentiful supplies of European sprats that would keep a population of young sharks well fed. Young of the year have also been found farther to the north near the entrance of the Çanakkale Strait, which leads to the Sea of Marmara and the Bosphorus.

Large, mature female white sharks have been caught in the area. One of the most impressive was a 5-metre (16-foot)-long female caught in a purse-seine net on the Foça coast of Turkey on the eastern side of the Aegean. It was March 1991, and the white shark was the largest ever seen in Turkish waters. Between 1991 and 2011, a further four mature white sharks were hauled out of the sea, but a female with embryos has yet to be captured here. The Foça shark, however, was just south of a possible nursery area, so she could have been heading there to give birth or could have just dropped her pups when she was caught. As the shark was not dissected scientifically, simply gutted, finned and decapitated, no record exists of whether she was pregnant or not. However, these catches indicate, perhaps, that white sharks are beginning to return to their old haunts. As they get older, the juvenile sharks move around the offshore and nearshore islands to the north and south of the Gulf of Edremit, before they probably embark on longer excursions elsewhere in the Mediterranean area.

It was generally thought that these Mediterranean white sharks had spilled in from the eastern North Atlantic population, but research into white shark DNA, by researchers from the University of Aberdeen, revealed that their closest living relatives are in Australia. One scenario is that a bunch of white sharks migrated, like Nicole, across the Indian Ocean from Australia to South Africa about 450,000 years ago. They were caught up in powerful Agulhas rings that had formed with higher sea levels in a warmer and more turbulent ocean than exists today. Unlike Nicole, who used the anticlockwise swirl to turn back to Australia, they were swept into the Atlantic Ocean and into the cold and wide northward-flowing Benguela Current, which bathes the western coasts of Africa. Programmed to head east, they found the continent of Africa was in the way, so they followed the coast towards the north until they reached a gap, and that gap was the Strait of Gibraltar, the entrance to the Mediterranean Sea. With no way out, as the Suez Canal didn't exist (and it wouldn't have helped anyway as the water in the canal is too warm and hypersaline), the sharks were trapped and have been there ever since.

Their relatives in the Antipodes follow a north–south migration pattern, winter and summer, similar to that on North America's east coast. In Australian waters, there are two main populations: one off Victoria, South Australia and Western Australia, and the other off New South Wales and Queensland, extending down to New Zealand and its sub-Antarctic islands and as far north and east as the waters of the southwest Pacific, including New Caledonia, Vanuatu and Tonga. The divide is the ancient granite mass of Wilsons Promontory overlooking the Bass Strait, although a few individuals do cross over to the opposite side.

Many white sharks from the southern and western population feed on New Zealand fur seals and Australian sea lions around the Neptune Islands and the Sir Joseph Banks group of islands, at the entrance to Spencer Gulf off South Australia. Dangerous Reef, four rocks in the latter archipelago, was where pioneering underwater cinematographers Ron and Valerie Taylor filmed the real-life sequences for *Jaws* in 1974. They had previously starred in the first documentary on white sharks, with the title *Blue Water, White Death*, which was filmed in the early 1970s. It was seen by Hollywood movie producers Richard Zanuck and David Brown, and Valerie told US writer Charles Thorp what happened next. The producers sent galley proofs of Peter Benchley's novel *Jaws* with the question: did they think it would make a good film? Realising there might be a role for them, they endorsed it enthusiastically...

and, indeed, there was a role for them. Before the main filming started in the USA, they were commissioned to obtain the live footage.

Rodney Fox, who was the victim of a serious white shark attack but lived to tell the tale, was bait wrangler, and a series of unexpected events enabled them to get some extraordinary footage. As Bruce – the animatronic shark – was supposed to be about 25 feet (7.6 metres) long and the sharks at Dangerous Reef were no more than 16 feet (5 metres), Ron and Valerie hired a small abalone boat, built a smaller than normal cage, and took on a smaller than average actor. The actor, it turned out, didn't like being with sharks and was reluctant to enter the cage, which was just as well. A large shark got entangled in its wires, thrashed about wildly and eventually destroyed the cage entirely, it sinking to the bottom in pieces; so much for a 'safety' cage, but the footage was gold dust. Spielberg rejigged the script to include the sequence, so every time you see a real shark in the movie, it's an American white shark played by an Australian white shark; and these sharks go on extensive migrations.

Sharks from the Spencer Gulf islands tend to head west. They cross the Great Australian Bight to the south of South Australia and when they reach the Cape Leeuwin Lighthouse, most turn right and head northwards up the coast of Western Australia. Female sharks tend to swim offshore, following the edge of the continental shelf, and when they reach the Indian Ocean, they drop down into deeper water off the shelf. Males hug the shallow waters closer to the coast, although they tend to be at least 10 kilometres (6 miles) from the shore and in waters at least 50 metres (160 feet) deep. However, the big questions are: where are they heading and why? This is difficult to answer because there's nothing obvious, except the humpback whale migration which sees the whales dropping their calves, along with courtship and mating, especially in Camden Sound in the Kimberley region of north-west Western Australia in winter. It's the biggest calving and nursery area for the largest humpback whale population on the planet, the afterbirth alone a draw for sharks, let alone dead calves and old-timers that didn't make it, but white sharks seem not to go that far north.

Farther south, the situation is far from clear. White sharks are present throughout the year on the south and lower west coasts of Western Australia, although they are most common in spring and early summer on the lower west coast itself, including Perth, Bunbury and Mandurah, and least likely to be present during late summer and autumn. Even though there are these seasonal patterns, the sharks head in all directions with no discernible coordination at all times of the year, so the notion that they are following the humpback migration is possibly a false one. They are also not coordinated with seasonal events, such as pupping at the expanding fur seal colonies in the region.

The eastern population moves north and south along the east coast – north in winter and south in summer. White sharks also circumnavigate Tasmania and cross the Tasman Sea to New Zealand and islands farther south, such as the Auckland Islands and Campbell Island. One of the most southerly records is from Macquarie Island, halfway to Antarctica, when a large female from the South Australia population crossed the Bass Strait, brushed the southern shores of Tasmania and then headed south. The attraction here is a healthy southern elephant seal rookery, as well as three species of fur seals, the occasional leopard seal, king penguins and the endemic royal penguin. Scientists working on the island sometimes see seals with bite marks, so they were not surprised to learn that they are visited by white sharks. The sea surface temperature here is between 3°C and 5°C (37–41°F), but the white shark's heat exchange system enables it to survive in these frigid waters that may be whipped up into mountainous seas by winds that regularly hit 160 km/h (100 mph). It's a remote and harsh environment but its proximity to where the nutrient-rich Southern Ocean meets the warmer south-west Pacific, resulting in mixing and upwellings, means there is abundant food for wildlife. It's one of the richest wildlife locations in the world and white sharks take full advantage.

By contrast, at the other end of the migration range, the eastern population of white sharks might pitch up in Tonga, in the South Pacific. Here, sea surface temperatures are at the other end of the scale – summer averages 28°C (82°F), while winter is still a warm 23°C (73°F). The attraction could be humpback whales giving birth.

The waters around New Zealand have white sharks all year round, the stay-at-homes being immature individuals. On either side of North Island are thought to be nursery sites but the focus of research has been the Chatham Islands and Stewart Island, the former about 800 kilometres (500 miles) to the east of South Island and home to fur seals and the latter just 30 kilometres (19 miles) off the southern tip of South Island, with New Zealand sea lions and fur seals.

Tagging research at Stewart Island has revealed that most

of the white sharks are immature, with two males to every female, and a third of males are mature. With such a large immature population the aggregation is unlikely to be related to courtship and mating, so the attraction must be food. The greatest number of white sharks occurs around one group of the Titi (Muttonbird) Islands off the north-east of Stewart Island, where many fur seals are probably a draw. This causes problems for the Rakiura Māori people during their annual harvest of muttonbird chicks.

The sharks remain in the vicinity of the island from late summer to early winter, spending up to five months in the region before heading for the sun for the rest of the year. Many individuals embark on these long migrations, with destinations similar to the eastern Australian white sharks. The New Zealand sharks head for the Great Barrier Reef on Australia's north-east coast, the Coral Sea, New Caledonia, Norfolk Island, Vanuatu, Fiji and Tonga, but the sharks from Stewart and Chatham take different routes. The Stewart Island sharks head north-west along the west coast of New Zealand, while the Chatham Islands individuals head directly north. There appears to be no crossover between populations at the two islands, but they tend to overlap at the tropical end of their migration. Three sharks, tagged with dorsal fin tags, illustrate what's going on.

The first was Nicolas Cage. He made a beeline for New Caledonia, with a more random wandering on his return. Pip settled first for New South Wales before hugging the east coast of Australia on her way to Queensland. The third shark, Caro, swam northwards across the central Tasman Sea to reach the Coral Sea and also paid a visit to Queensland.

On their return, these sharks are so precise in their navigation skills that they head straight for the places from which they set off several months before. During their journeys, they can travel up to 150 kilometres (90 miles) per day, and they dive repeatedly to depths of between 200 and 800 metres (650–2,600 feet), one individual making a record dive to 1,246 metres (4,088 feet). With such a diving regime, they experience a wide range of sea temperatures from 27°C (81°F) at the surface in the subtropics to 3°C (37°F) at depth. Why they go down so deep is a bit of a mystery, although it has been suggested that they find deep-sea squid and fish to eat. Another reason might be that they seek out north–south running seafloor ridges with magnetic identities that they use for navigation, but we've yet to understand precisely how it all works. Let's face it, there's still a large amount of white shark biology with which we have to come to terms.

CHAPTER 7

SENSING AND SOCIALISING

White sharks are one of the few species of shark to lift their head out of the water and look around. It's a behaviour known as spyhopping and more usually seen in marine mammals, such as orcas and humpback whales, and it's sure to impress. People in boats having a close encounter with a white shark come away with the distinct impression of being sized up as potential prey. The shark might inspect each person on board one by one. It's decidedly spooky, and reminiscent of that classic Martin Brody moment in *Jaws* when he gets his first close look at the shark. He backs into the wheelhouse and says to Captain Quint, 'You're going to need a bigger boat!' I think most people who have had the privilege of being close to a white shark will have had the same thought. When a large shark raises its head close by and looks you in the eye, it's a heart-stopping moment.

During spyhopping, seawater is sometimes seen to squirt or dribble from the eyes. It's thought this is caused by the shark squeezing its ocular muscles to push the eyeball against the back of the eye socket in order to change the shape of the lens. This helps it see better in the air, as air at sea level is 784 times less dense than seawater. It's actually a similar system by which human eyes work, except less sophisticated. The result is the white shark can see in air, but the image is not very sharp.

Vision, however, is one of the white shark's principal senses, as it is mainly a diurnal predator, although not averse to hunting at night; in fact, a study at Guadalupe Island, off the Pacific coast of Mexico, showed that some individuals actually prefer to hunt at night. The majority, however, hunt during the day, during which the white shark relies heavily on vision, especially at close-range distances within 15 metres (50 feet), and has much more success when the target is at the surface. Its search image is the silhouette of prey, such as seals, against the background of light at the surface. The structure of its eyes enhances this. The retina is divided into two sections, one for day vision and the other for night vision, and contains rods and cones. The cone photoreceptors in the retina are responsible for higher spatial acuity and there is a retinal region – the area centralis – for acute vision that samples a region of the visual field in a zone above and to the sides of the head, so it's well equipped to look up. With eyes on either side of its head, the white shark also has almost a 360-degree general field of vision, but there are two blind spots: directly in front of the snout and directly behind the head.

Additionally, the white shark cannot see colour and can't perceive detail as well as the human eye can. With this in mind researchers at Australia's Macquarie University have been looking at the way in which young white sharks, switching from fish to seals, acquire and use a search image for their new prey. This is combined with other senses, such as smell and taste, to know what's good to eat and what's not. It's a learning process that inevitably leads to mistakes.

The experiment took the shark's viewpoint looking up towards the surface and tried to determine what motions and shapes the shark would see. The researchers compared a seal swimming, a human swimming and humans on short surfboards (1.8 metres/6 feet) and long surfboards (2.8 metres/9 feet), and they found that there was little to choose between them. None would be visually distinct to, say, a juvenile white shark swimming below. The motions and shapes of all the human scenarios were similar to those of a seal, although there was a little discrepancy between seals swimming and longboard surfers paddling at the surface, but as long surfboarders get hit as often as short surfboarders, it seems the sharks don't notice the difference. The experiment shows that most shark bite incidents on humans probably were cases of mistaken identity by younger, inexperienced sharks. However, one other thing they noted is that mature white sharks, which are the more experienced hunters and tend not to make mistakes, do bite people. It means that some attacks on humans were possibly not mistakes. The sharks knew what they were doing. They deliberately targeted humans as food. Maybe they were simply hungry.

During an attack, the white shark has a unique way to protect its eyes from a seal's flailing claws. While most other species of shark have a nictitating membrane that slides across the eye, the white shark swivels its eyes back into its sockets, which means that for those last few seconds before making contact, it is effectively swimming blind; that's when a second sense kicks in, one that is unique to sharks, their relatives and a handful of other creatures – the electromagnetic sense.

Like most other shark species, the white shark has numerous small pores dotted around the snout, known as the ampullae of Lorenzini. Sharks in general have an average of 1,500 of them, although hammerheads have over 3,000. Some dedicated souls counted the pores on a black reef shark and found there were more on the right side of the head than the left, which is why this species tends to take hand-held baits with a right sideways snap. Whether white sharks have a similar imbalance is not known, as nobody appears to have done the counting. The ampullae were first described by the Italian biologist and physician Marcello Malpighi (1628–94), of Malpighi corpuscles in the kidney fame, but the first detailed account came in 1679 from another Italian physician, Stefano Lorenzini (b. 1652). At first, nobody could work out the function of the pits. They were clearly sensory in nature, but exactly what they sense eluded these seventeenth-century scientists. It was found that they had a response to slight changes in temperatures and to weak mechanical stimuli, but it wasn't until as recently as 1960 that R. W. Murray at Birmingham University revealed that their true function is to detect electrical fields. Contained within each pit is a highly conductive jelly and the pits are linked together by jelly-filled canals below the skin. Inside are receptors that can detect infinitesimally small electrical currents that are generated, for example, by the heart muscles of the prey. The jelly, which is clear, contains keratan sulphate, one of the most conductive biological substances. So, when the shark is close to its target and is swimming blind, the ampullae give an indication of precisely where it should bite. However, sometimes things can go wrong. A small boat with an outboard motor will create an electrical field in the water from the metal of the engine. The shark, believing it to have found the electrical activity of its prey, will start to mouth and even attack the engine. It looks as if the shark is attacking the boat and the people in it, especially if it previously put its head out of the water and scrutinised the occupants, but in reality, it's suffering from sensory overload and is thoroughly confused.

These sensory pits can detect variations of 5 billionths of a volt per centimetre, the equivalent of two AA batteries connected 16,000 kilometres (10,000 miles) apart, but electric fields are not the only phenomena that they pick up. They can also sense variations in the Earth's geomagnetic field and detect the magnetic signatures of rocks, so they function like a natural GPS system which helps guide the shark on its migrations.

The ampullae evolved more than 430 million years ago from the mechanosensory lateral line systems of early vertebrates. This system is present in modern sharks as a series of fluid-filled canals just below the skin. They run along the flanks and over the head and stretch to the tail. They are open to the sea via a string of visible sensory pits, quite distinct from the ampullae of Lorenzini, and so are able to detect movements, vibrations and changes in pressure in the surrounding seawater, including the water passing over the shark's body. This gives the shark spatial awareness, an ability to navigate and an indication of how fast it's going, and despite being a passive system, it is thought to be able to discriminate between objects in the sea by their shape and to pick up the movements of prey. Pressure waves from its own movements are also thought to bounce back from objects in its environment, much like the sonar other animals use, and so it is able to build up a three-dimensional map of where it happens to be and be aware of other creatures moving through the field.

The lateral line was documented first by several seventeenth-century biologists – the Danish pioneering anatomist Niels Steensen (1638–86) in 1664, Lorenzini in 1678 and the German physician and botanist Augustus Quirinus Rivinus (1652–1723) in 1687. At first, the 'lateral canals', as they were known, were thought to be where fish produce mucus, and that view prevailed until the mid-nineteenth century. In 1850, however, the German anatomist Franz Leydig revealed that the canals contained sense organs, and since then a whole army of biologists from all over the world has been trying to work out precisely how the lateral line works and the detail of its various functions. Suffice to say, the system has been described as 'touch-at-a-distance'.

For really distant sensing, two more sensory systems fill the role – hearing and smell. A white shark's ears are not very obvious: just a tiny opening behind and above each eye, but what they lack in size, they more than compensate for with sensory power. The tiny sensory cells inside can detect sounds, especially low-frequency sounds between 20 and 300 hertz, that travel faster and further in water than in air. Depending on the amplitude and distance of the source, in tests they have been shown to be able to pick up those sounds from more than a kilometre (0.6 miles) away. Low-pitched sounds of an animal struggling, which are generally irregular and below 40 hertz, for example, are like a dinner bell to sharks, and white sharks are no exception. They also have an 'ear stone', a kind of balancing organ that indicates their orientation in the water, whether nose up, nose down, the right way up or upside down.

A lot of rubbish has been written about a shark's sense of smell. Some commentators have claimed that a shark, species unknown, can detect a tiny drop of fish oil in an Olympic-sized swimming pool, but much of this must be taken with a pinch of salt. Sharks certainly have a keen sense of smell, far better than ours, but they cannot smell anything like a drop in an Olympic pool, according to researchers from the Florida Atlantic University. They indicate that the smallest quantity that sharks can smell is one drop of scent in a billion drops of water, the equivalent of one drop in a back garden, family-sized swimming pool, and there is a fundamental reason why. This is roughly the background concentration of amino acids dissolved in coastal waters. If they had evolved the ability to detect smaller quantities, it would have been of no use as they would not be able to tell the difference between a potential food source and all the random chemicals in the sea. Sharks in the open ocean might be able to detect a little less, compared to coastal species, as the background 'noise' is less.

To pick up smells the white shark has two nostrils or nares underneath the snout. As the shark swims, water passes through one side of the nostril, then flows through a nasal sac and finally out the other side. Inside the sac are folds of skin – the olfactory lamellae – which greatly increase the surface area, enabling the shark to better pick up smell molecules. Highly sensitive receptors in the nasal sac can detect incredibly small concentrations of chemicals and nerve signals are sent to the brain, in particular the olfactory bulb in the fore-brain. The white shark's is the largest, relative to body size, of any shark. About 19 per cent of a white shark's brain mass is dedicated to smell, an indication that olfaction is probably as important a sense as vision, used in finding not only food but also a partner.

The sense is also directional. With two nasal cavities, any smell coming from, say, the right, will reach the right nares momentarily before the left. By sweeping the head slightly from side to side in the smell corridor, the shark can trace the smell back to its source.

On occasions, white sharks can be fussy eaters. Taste receptors on the tongue, on the lining of the mouth and in the pharynx enable a shark to decide whether prey is worth eating or spitting out. They often spit us out. They also have particularly sensitive teeth and gums with pressure-sensitive nerves and so will mouth objects to check them out. Aside from anything with blubber or meat, they've been known to mouth buoys, driftwood and boats, and it's not necessarily a gentle affair. In its January 1983 monthly report, the California Department of Fish and Game recorded three instances of crab-trap buoys being mouthed by white sharks. Tooth impressions, scrape marks and teeth embedded in the cork indicate the mouthing was carried out with gusto. Similarly, fishing gear associated with salmon trawling in south-west Oregon was mouthed by a white shark estimated to be 3–4 metres (10–13 feet) long. The tooth impressions in the trawl's otter boards are testament to forceful bites. Having checked out the gear, the shark simply disappeared. As sharks lack hands, as you've already realised, they use their teeth instead. It reinforces the notion that white sharks sometimes look as if they're attacking when actually they're exploring.

To digest all of this sensory information, the white shark has a well-developed brain. It's possible that white sharks are more brain than brawn. Researchers consider them to be intelligent and to learn rapidly, and they are surprisingly inquisitive.

There's no question that dolphins and seals are intelligent

mammals, so the white shark has to be pretty clever to outwit them; in fact, the white shark is as intelligent as it needs to be to survive in its environment and that intelligence grows with experience. Young sharks are often quick to bite, moving in where angels fear to tread, but as they get older and wiser, they might stand off and use their brain power to select the most effective, safe and energy-efficient way to tackle their prey – strategic planning. Nevertheless, only 50 per cent of strikes are thought to be successful.

Around Guadalupe Island, one of the cage diving companies, Nautilus Adventures (at the time of writing denied access because of the Mexican government's ban on tourism here; see also Chapter 12), has been monitoring white sharks – and some pretty big ones by all accounts – as they come in and try to take baits placed on the sea's surface. On board the boat, bait wranglers entice the sharks close to the cages but whip the bait away at the last moment to keep the sharks interested. What they've noticed is that the sharks appear to assess the skills or otherwise of the wrangler, recognise their weak points and then 'capitalise on the exact moment the wrangler takes his eyes off the bait'. And even before they chase a bait, they swim around the boat to check things out, taking account of the position of the sun and the reflections on the water that could make the wrangler's job more difficult in knowing where the shark is going to appear. It's all pretty clever for a fish, and the white shark – let's be agreed – is a pretty clever fish.

Biologists say that social animals are more intelligent than solitary ones, so how come white sharks appear to be the brainiest species of shark? For centuries, it was thought that white sharks were complete loners, except, of course, at times of procreation, but modern observational science has revealed that they can be social creatures after all.

So, what's this all about? Social interactions occur when the behaviour of one individual affects the behaviour of another from the same species. These dealings can manifest themselves in all sorts of ways: aggressive behaviour between individuals, cooperation in hunting, learning from others' success or failures, reproduction, gathering together in clans for mutual benefit and simply as travelling companions, and white sharks appear to exhibit many of these types of social behaviour.

In the north-west Atlantic, for example, the OCEARCH project found two juvenile white sharks that appeared to travel together as they headed northwards on their spring migration. They had been given the nicknames Simon and Jekyll, and they were almost inseparable. In December, they were first tagged off Georgia, and by March they were both off North Carolina. There was a brief parting of the ways in May when Jekyll was off New Jersey and Simon had carried on to Long Island, but by early July they were back together again off the coast of Nova Scotia and by the end of the month in the Gulf of St Lawrence, having travelled 6,400 kilometres (4,000 miles). In the eastern North Pacific, off the Farallon Islands, pairs of white sharks have been seen to follow each other when hunting seals, and at South Africa's Seal Island, stable clans, with between two and six individuals, arrive and depart together each year.

Researchers from Florida International University have observed the white sharks that arrive seasonally at Guadalupe Island. They found that the sharks gathered understandably close to seal and sea lion breeding colonies, but there was more to it than that. The sharks showed several social associations. Generally, male sharks hung out with males and females with females, but there was a lot of variation, and the intensity of those relationships varied from shark to shark. One of the more gregarious sharks socialised with twelve others during the course of thirty hours, while a less socially active individual took up with just two other sharks in five days. Associations, however, could last for unexpectedly long times. One shark remained close to another for fully seventy minutes, so it is unlikely to have been a random event.

Although they might hunt in close proximity to others, generally it's not what scientists would normally consider coordinated, cooperative hunting. The sharks are simply exploiting each other's strengths and maybe learning from the successes or failures of others, a more remote form of cooperation. Even so, shark biologist Leonard Compagno tells how fishermen off South Africa have seen two sharks working closely together. One distracts a seal from the front while the other sneaks around behind and grabs the prize. They might share the spoils, but with a caveat: the bigger shark gets the lion's share and woe betide any lesser mortal that should muscle in.

In encounters with individuals that are smaller and therefore lower in the pecking order, any small shark that oversteps the mark will be put firmly in its place. It will be on the receiving end of assertive, and really quite violent, body slams, particularly in the area of the gill slits, a shark weak point. Many sharks show damage there, the result probably of hostilities when they were somewhat smaller and on the receiving end of an admonishment. Jon Capela possibly had the same experience in April 1989.

Capela and some friends were aboard his 10.7-metre (35-foot)-long boat about 1.5 kilometres (0.9 miles) south of Point Pinos in Monterey Bay, when they came across a white shark devouring a common or harbour seal. Jon estimated the shark to be 5–7 metres (16–23 feet) long, so quite a large animal. They stopped to watch, only for the shark to leave its meal and circle the boat, which it then rammed in the bow four times within twenty minutes. As a last act, it slammed its tail down against the swim step and propellor on the stern, before returning to the seal. Interestingly, the boat was painted white and blue, the colour white being significant. White surf boards, for instance, are hit more than those in other colours. It is quite possible, even probable, that the large shark had seen the boat as a rival; after all, it had the right colours, and the fact that it seemed to be taking an interest in the shark's food more than likely aggravated the attack.

A similar event occurred some decades before, off the coast of Cape Breton Island, Nova Scotia, a known northerly summer hotspot for white sharks. It was in July 1953, when two commercial fishermen went out in their dory to collect lobsters. There were always many dories out fishing, but this particular dory was the only one painted white. It was the one that stuck out from a shark's point of view, resembling the colours of a rival. And, lo and behold, every day for the best part of a week, the white dory was closely followed by a white shark. The other fishermen looked on in disbelief, but then things took a turn for the worse. One day the white dory was out alone, and at this moment the shark decided to attack. It rammed the boat, creating a large hole in the bottom and with such force that the fishermen were tossed into the sea.

There was a considerable chop, and one didn't make it, while the other clung to the remains of the boat until help arrived. Examination of the hull revealed a tooth fragment, which was identified as coming from a white shark about 4 metres (13 feet) long. Why had the shark attacked? Did it see the white dory as a rival? We'll never know, but it's more than coincidence that a white boat and a white shark had a violent clash.

This rather vicious form of communication is a mechanism whereby sharks can maintain a hierarchy, a type of social interaction whereby members of the society are ranked according to relative status, and in shark society size counts. It's a way to keep the peace, as long as individuals know their place, but in the Gansbaai area of South Africa, many of the white sharks in the local visiting population are roughly the same size, yet they are not constantly fighting for supremacy. There must be some other factors keeping them from coming to blows. A team of Italian researchers tried to find out how they behave towards each other by observing pairs of white sharks that came to wrangled baits on cage diving excursions between 2009 and 2018 (with fewer sightings in 2017 and 2018 when killer whales scared away many of the sharks). The results were subtle but quite revealing.

When pairs of sharks visited the baits, they showed a total of eight different social interactions. The most common was known as a 'swim by', during which two sharks swam towards and passed each other, so were not on a collision course. They did, however, remain parallel as they passed by, the distance between them varying between 0.5 and 2.5 metres (1.6–8 feet). In a 'parallel swim' two sharks swam close to one another in the same direction and remained parallel, a distance of between 0.5 and 2 metres (1.6–6.6 feet) separating them. The behaviour ended when the shark slightly behind peeled away. A third manoeuvre was the 'follow/give way' in which one shark closely followed another, causing the leading shark to turn away. A fourth was simply the 'follow'. One shark followed another and copied its movements, the following shark often with its mouth open.

Mouth gaping is thought to be another form of intraspecific communication. What the shark is saying to another shark is unclear, but they also gape when close to cage diving boats. Having had a bait whipped away in front of its nose, the shark turns slightly on its side, its head comes above the surface, and it opens and closes its mouth, and so gapes several times for about ten seconds, what Wesley Strong of the Cousteau Society has described as 'repetitive aerial gaping'. It's as if the shark is frustrated by the thwarted attempts to catch the bait and dissipates it by gaping, while at the same time serving to reduce the intensity of any intraspecific conflict; in other words, the shark is less likely to take out its frustration on its neighbours.

A fifth behaviour was the 'give way'. Two sharks swam towards each other on a collision course and, at the last minute, one turned at right angles to the left or right. The shark that did not deviate must have been somehow dominant over the other, which led to the sixth event, the 'stand back', when two sharks headed towards each other and they both turned away, so a dominance hierarchy cannot have been established. An unusual behaviour only witnessed once was the 'piggy back', during which one shark swam down to the back of another and they continued along together for several seconds before parting. The individuals involved were male and female so there must have been something of a sexual nature going on. The eighth event was a 'splash fight'. One shark splashed another with its tail, and the other would repay it in kind. It's something that has been seen also during feeding events off the Farallon Islands: a shark turns on its side, raises its tail and then brings it down with some force on the sea's surface with a loud thwack. The display is generally aimed at another shark invading the performer's patch. Similarly, the shark might breach, pushing about of a third of its body out of the water at an angle of between 30 and 60 degrees and then slam back down in a huge splash, much like a humpback whale. It's thought to be a severe warning to another shark trying to feed and therefore competing in the same area.

Another sign that a shark is unhappy about the presence of another is a subtle agonistic display. It's best seen in the more flamboyant grey reef shark, which bends its body into an S-shape, pushes its pectoral fins straight down and swims erratically. Should the intruder not back down, the shark bears down with mouth agape and makes a slashing attack with its teeth. The white shark has been seen to perform a similar, but less dramatic, display.

One thing to bear in mind, however, is that not all white sharks do all these things to the same extent. Individual white sharks are like individual people. They're not automatons, as many people think. Some are more curious than others, and others are more belligerent, some assertive, others timid, so how a shark behaves towards other white sharks, seals, whales and people depends on the shark. They're not all the same. They're like us.

CHAPTER 8

HEALTHY APPETITE

White sharks and seals have lived alongside each other for over 6 million years. The sharks have hunted the seals during that time, and the seals have evolved ways to get away. There is even fossil evidence to suggest that this was, indeed, the nature of their relationship right from the beginning. A fossil tooth from an early Pliocene white shark was found embedded in a fossil seal's rear flipper. There were no signs of healing tissue, indicating that the seal either lost its flipper during the encounter or didn't survive. This two-for-one fossil had been locked away in sediments about 4.5 million years ago, and they were laid down in what is now Florida. The ongoing relationship between the two animals today, though, is more easily seen on both coasts of North America – off California and Massachusetts – and white shark hotspots elsewhere in the world, such as South Africa and South Australia. In all these places, white sharks stake out the rookeries and haul out sites of seals and sea lions.

On the west coast of the USA, two notable sites are the Farallon Islands to the west of San Francisco and Año Nuevo on the mainland to the south. Año Nuevo is where early work on northern elephant seals, one of the white shark's main prey items, was carried out by scientists from the University of

California, Santa Cruz, and the Farallons is where much of the pioneering research on white sharks and their interaction with seals was conducted by scientists from the Point Reyes Bird Observatory. A couple of them switched from seabirds, of which there are many colonies on the Farallons, to white sharks and seals. In fact, the archipelago has a smorgasbord of pinnipeds from which the sharks can choose – male, female and young northern elephant seals, California sea lions and harbour seals – and the sharks stake out the haul-out sites at times of the year when the most vulnerable seals are present.

Aside from the regular visiting seals that come to mate, give birth or to moult, other young seals from rookeries to the south, such as those from southern California, pitch up at the Farallon Islands. These young seals are quite unaware of the geography of the islands and threats that are present, so this is precisely the kind of naïve and unaware prey that the sharks have been waiting for, the ones easiest to catch. Indeed, the sharks seem to have a preference for these novices, appearing close to the haul-out sites when they are mainly present but moving away when the prey have also left the area. Faced with such a glut, some individuals have become unexpected creatures of habit. One shark, for example, was recorded having visited the same northern elephant seal rookery at the same time of year for twenty-six years, where it snacked almost solely on juveniles. So abundant is the food supply that another shark was seen to catch three juvenile seals during the course of a single week.

At the Farallons, most of these attacks are within an area relatively close to the shore, a so-called 'killing zone' that surrounds each of the islands in the archipelago, and the sharks do not hunt aimlessly, hoping to bump into a seal. They have a distinct search pattern during which they zig-zag back and forth just offshore and parallel to the shoreline. The smaller, inexperienced sharks cover a greater area than the larger, more experienced individuals. These old hands focus on areas where they have had success in the past. They also keep their neighbours in view and maybe note their successes or failures.

The sharks avoid the surface, remaining closer to the bottom, a tactic adopted by a predator that has a silhouette-based hunting strategy. It can look up and see its prey backlit by the light at the surface. Looking down, the colour of the shark's body renders it effectively invisible to the seal. The dark back and white belly is a form of countershading that means the shark is camouflaged against the rocks, seaweeds or the deep blue of the ocean when seen from above, while viewed from below its belly blends in with the sunlight from the surface. Off the Farallon Islands, white sharks have been seen to avoid the sandy sections of seabed, which would give them away, and so they remain over dark, rocky, seaweed-encrusted reefs in a bid to remain unseen. This is helped by the fact that melanocytes in the shark's skin can render it darker or lighter depending on the background.

They try to stay below and behind their target, using the position of the sun strategically. When approaching a target, the shark ensures that the sun is directly behind it. According to researchers from Flinders University, 'on sunny days, sharks reverse their direction along an east–west axis from morning to afternoon', but if it is overcast, they approach from any direction. If undetected they are ready for the sudden last-minute assault, when they might rush almost vertically upwards at full throttle and slam into their victim without it even seeing them coming. Well, that's the theory, but off the coast of South Africa, the Cape fur seals are wising up to this kind of attack. At the last moment, they jig to one side or leap clear of the water, causing their pursuer to overshoot and also leave the water, sometimes entirely, to crash back down in a fountain of spray.

Nevertheless, off the rookery on Seal Island, the sharks successfully catch a fair number of the seals. They target lone young-of-the-year youngsters moving between their rookeries and feeding grounds 65 kilometres (40 miles) out at sea. The seals swim at or near the surface and must run a gauntlet of white sharks. In one study led by the University of Vancouver, between 1997 and 2003, 2,088 predation events were recorded, each lasting less than a minute. The sharks bite obliquely, using their anterolateral teeth and with a rapid lateral snap of the jaws, rather than vertically as was first thought. This enables them to make the biggest possible wounds.

The best time to hunt seals at Seal Island, apparently, is during the hour after sunrise, with successful predation reducing as things brighten up. When the success rate drops to about 40 per cent, the sharks pack it in for the day. It's the first time the cessation of predation when conditions are not optimal has been recorded in any species of predatory fish. Also, hunting success is significantly less in areas where the frequency of predation is highest, where overcrowding has an impact. It's as if any predatory attempt here is half-hearted, another first – the first time social factors have influenced predatory behaviour in sharks. So, if you want to witness 'Air Jaws', as the breaching sharks became known during Discovery Channel's Shark Week, you'd better get up early.

An international team of scientists led by researchers from the University of Tasmania used on-board cameras, accelerometers and depth sensors to track how the white shark achieves such feats. Its attack starts about 20 metres (65 feet) down, when the shark accelerates with a 6.5-fold increase in tail beat frequency and speed, and heads rapidly for the surface with a change of pitch of about 30 degrees. The ascent lasts no more than 16 seconds, during which the shark is able to make rapid adjustments to its trajectory. It then either slams into the seal and the two leave the surface briefly or the seal jigs and the shark flies high into the air having missed. Similar behaviour photographed from a drone in California shows white sharks breaching while not feeding. Precisely what they are doing is far from clear.

When feeding, exactly where the shark makes contact with its prey depends on whether the pinniped is a seal or a sea lion. Sea lions and fur seals – the so-called 'eared seals' – propel themselves along with their front flippers. The main propulsive force for harbour seals, grey seals and elephant seals comes from their back flippers. White sharks are surprisingly observant: they know which is which. They hit seals abaft and sea lions to port and starboard. With its main propulsive force out of action, the prey cannot escape. Now, that's pretty clever for a fish.

If successful in their ambush, the sharks appear to behave differently depending on the size of the prey and its ability to fight back. Large northern elephant seals, for example, are treated with caution. The seal is immobilised by a debilitating bite to the rear flippers, after which the shark sits back at a safe distance, out of the reach of flailing claws, and waits for its target to bleed to death before beginning the feast.

Smaller prey, such as the more frequently encountered seals and sea lions, are ripped apart with little hesitation, although the sharks are still careful to avoid being injured by the seals' claws. They'll demolish a California sea lion in just five or six enormous mouthfuls, and an extremely naïve young grey seal, which was tracked by a drone at Cape Cod, swam straight into a white shark's mouth and was finished off with a bout of head shaking and two enormous gulps.

Although it might not be too obvious when they take apart prey, white sharks are actually rather delicate eaters. They take care not to damage their teeth, by making sure that they don't bite down too hard. That could result in the teeth breaking on the victim's large bones. Even so, the white shark has a bite force of 1.8 tonnes, three times that of an African lion and twenty times more than a human. Despite their reticence to tackle bone, they can, in fact, munch through a victim's limbs with a bite that has been claimed to be the strongest of any living animal.

Teeth in the lower jaw are more pointed and act like forks, while the triangular, serrated teeth in the upper jaw are more like steak knives. When taking a bite, the shark shakes its head from side to side so that the serrated teeth literally saw off a chunk of blubber. The telltale sign that a white shark has been feeding is a semicircular wound where the shark has removed the skin and any fat and meat underneath. Should they be clumsy and break a tooth or two, there are plenty more to take their place, about fifty working or active teeth at the front of the mouth and another five or six rows of replacements waiting behind in case they are needed. A long-lived shark might produce as many as 30,000 teeth during its lifetime.

White sharks, however, do not have everything their own way. They rely on surprise, so if their cover is blown, they usually withdraw and hunt elsewhere. Off the coast of South Africa, Cape fur seals see them on their way, thus gaining the upper hand. If a white shark is spotted, for example, and the element of surprise is lost, the seals mob the shark, like crows mobbing an eagle. Instead of swimming at high speed away from the shark, they move towards it and harass it until it is chased away from their rookery.

One thing that shark watchers in South Africa have noticed is that, on most of their expeditions, they see only the younger adult sharks, the seal hunters. The large, mature females are noticeable by their absence but should the carcass of a large whale – southern right, humpback or Bryde's whale – float in on the tide, the giants suddenly appear. They're done with whizzing all over the place in pursuit of something that's likely to get away and have opted for the easy life, so the largest white sharks have a tendency to home in on the bloated bodies; after all, they don't try to get away and there's a substantial amount of high-energy blubber available just for the taking. The sharks are able to slice off their large semi-circular bites without interference, even from others of their own kind. Sharks queue for the right to have a bite, the smallest sharks deferring to their larger relatives, or they get attacked in their gill area for their trouble.

Having taken a few large mouthfuls, however, the sharks become surprisingly soporific. They appear to be in a trance; indeed, they are so out of it that people can swim close to them with impunity. There is film, for example, of free divers in Hawaii swimming right alongside an enormous female white shark by the name of Haole Girl, without being attacked, and cage divers off Mexico's island of Guadalupe have been able to interface with another giant female, the well-known Deep Blue – star of Discovery Channel's Shark Week – without being in danger. In Hawaii, both Haole Girl and Deep Blue have been seen within days of each other gorging on a sperm whale carcass, having frightened away all the tiger sharks that would usually do the cleaning up. Both these sharks are giants, close to 6.5 metres (21 feet) long – two of the largest living white sharks.

Off the east coast of Brazil lies the Abrolhos Bank, a biodiversity hotspot, a destination for whales – especially humpback whales – and a stop-over for white sharks on migration. The whales come here to breed and there are many carcasses going for the taking. In one study by several Brazilian universities and institutions between 2001 and 2010, of 150 humpback whale carcasses, thirty-five showed signs of being fed on by white sharks, and three living whales also had white shark bite marks.

Indeed, it has become apparent that these larger sharks don't stop at dead whales. They have a go at live ones too. Off the coast of Western Australia scuba divers watched as a large white shark harassed a humpback whale mother and her calf, while a male whale slammed its tail flukes down on the sea's surface trying to protect them. The two most extraordinary observations, however, were made off the coast of South Africa.

One was from a research ship in Mossel Bay in February 2017. A 7-metre (23-foot)-long humpback whale had been weakened, having been caught in a fishing net, when along came two white sharks. The first to arrive was 3.5 metres (11 feet) long and proceeded to circle the whale in what looked like a visual inspection, before taking a 'test bite' behind the pectoral flippers. This caused copious amounts of blood to spill into the water. After about forty minutes, it made a second bite into the peduncle – the thin section between the tail flukes and the whale's body. More blood was lost, but on both occasions, the shark did not remove any blubber. Shortly afterwards, a second shark appeared. It was estimated to be about 4 metres (13 feet) long. The smaller shark abruptly disappeared, possibly intimidated by its larger companion, whereupon the second shark also bit into the tail stock causing even more bleeding and producing a red slick that could be seen to extend 50 metres (55 yards) away from the whale. After more circling, the larger shark slammed into the whale's rostrum (snout) with its mouth agape as if taking a test bite. After attacking the peduncle again, about ninety minutes into the encounter, the whale died and sank. The larger shark continued to circle on the surface, and that was how it was when the research ship moved on.

The second event took place off Port Elizabeth, about three years later, when a drone found a 4-metre (13-foot)-long white shark attacking a 10-metre (33-foot)-long humpback whale. The shark was thought to be a female called Helen, and by all accounts, she was especially ruthless in the way she went about her attack. Like the other sharks, she bit into the peduncle to cause bleeding and after enormous quantities of blood seeped into the sea, she grabbed the whale by the head and tried to drown it. After fifty minutes, it succumbed to its injuries. These were two unusual white shark bite events in which the individuals had been surprisingly 'strategic' and showed just how resourceful white sharks are.

While the floating carcasses of giant baleen whales are a natural draw for white sharks, other cetaceans (whales and dolphins) are also on the menu. In the Mediterranean, for example, dolphins are a main food source, replacing monk seals, which are now rare and on the brink of extinction, and off the coast of Sarasota, Florida, about one-third of dolphins have scars caused by sharks, while in Shark Bay, Western Australia, the figure is closer to 70 per cent. In Hawaii, spinner dolphins are regularly attacked by sharks, white sharks among them, and a study at the University of Queensland revealed that nearly 40 per cent of bottlenose dolphins seen in in Moreton Bay, Queensland, have wounds that were the result of shark attacks, mostly from white sharks and tiger sharks.

The sharks, approaching from behind, either from underneath or on top, cannot be detected by the dolphins' sonar, which scans the sea ahead of them. Directly behind is a blind spot. Again, like the attacks on seals, the sharks bite into the tail region and so immobilise the main propulsive force or they take an enormous chunk from the flanks, which serves the same purpose in bringing the dolphin to a halt. When coming from above, they will bite into the dorsal area, just ahead of the dolphin's dorsal fin, as well as just in front of the tail. The shark, though, might not retain its prey.

In August 2018, oyster fishermen in Smoky Bay, on the west coast of the Eyre Peninsula, South Australia, witnessed a 3-metre (10-foot)-long white shark kill a young bottlenose dolphin and then circle the body for some time. The delay in devouring its victim was its downfall. A larger white shark, estimated to be 3.7 metres (12 feet) long, swam in aggressively and frightened away the first attacker, stealing its prize. Even a size difference of as little as 70 centimetres (28 inches) counts for something in the white shark world.

The dolphin attackers, however, are not necessarily the larger sharks. Over the last weekend of May 2023, a vicious bunch of juvenile white sharks was filmed tearing apart a dolphin just off Torrey Pines State Beach, near San Diego. Why the dolphin was alone and not with its pod is hard to tell, as the group would probably have been able to see off the sharks, but whatever the reason, it was a big mistake. A woman had been bitten previously in the same area, not seriously, probably by a juvenile white shark, so these pesky youngsters are not the sweet innocents that you might imagine.

Generally, if dolphins spot a shark tracking them, the boot is on the other foot. Like the fur seals of South Africa, a pod of dolphins will gang up on the shark and mob it, sometimes ramming it in the area of the gills or the soft underbelly with their hard and bony rostrum, a biological battering ram, the gill slits being a particularly sensitive area for sharks of all types, not only white sharks. Generally, though, the dolphins do not strike sharks. Off the coast of South Africa, for example, a pod of humpback dolphins was seen to chase a 4–5-metre (13–16-foot)-long white shark, a known dolphin killer. Although there was no physical contact between the two species, the dolphins pushed the shark out of the area. In October 2022, in South Africa's Plettenberg Bay, on the coast of Western Cape Province, another pod of dolphins swam directly at a 3-metre (10-foot) white shark and surrounded it, one dolphin swimming directly over the shark's back to emphasise they meant business. A shark had killed a human swimmer the previous month in Plettenberg Bay, but the dolphins were having none of it. Again, they chased the shark out of the bay to the open sea. Dolphins are more flexible and manoeuvrable than white sharks, which gives them a distinct advantage.

While dolphins, such as the bottlenose dolphin, are something of a nuisance without causing serious harm, there is one marine creature that even large sharks fear. The world's largest living dolphin – the orca or killer whale – is the white shark's nemesis. Size for size sharks and orcas are a fairly even match. Adult female orcas, which do all the hunting, are 5–7 metres (16–23 feet) long, slightly larger than the sharks, which are in the 4–5-metre (13–16-foot) range, but the mammals have a distinct advantage: they hunt cooperatively and appear to be more intelligent.

The first time that violent encounters between white sharks and orcas came to light was off the Farallons in 1997. Fishermen off South Farallon Island witnessed two orcas chowing down on an elephant seal, when a young white shark came along and tried to muscle in. The whales turned on him, rammed him repeatedly and eventually killed him. They then ate his oil-rich liver. Somehow, word must have got out because each time orcas appeared in subsequent years, particularly in 2009, 2011 and 2013, most of the white sharks in the area skedaddled and didn't come back for the rest of the season. During those years, white shark attacks on pinnipeds dropped by more than 60 per cent.

In the aftermath of what turned out to be an orca attack, a white shark's body washed ashore in South Australia in October 2023. The head and tail were intact, but all its innards were missing. Scientists were able to confirm that orcas had done the damage because when they took swabs for DNA analysis, an orca's signature was all over the shark's body.

Similar events occurred off the coast of South Africa. The culprits this time were what are known as 'flat-toothed' killer whales. They are shark specialists, and they get their name from wearing down their teeth on the sharks' abrasive skin. Scientists noticed these shark-killing orcas in 2015 and 2016 when many broad-nosed seven-gill sharks were killed at a known aggregation site in False Bay, south of Cape Town. All were missing their liver. Then, in 2017, the action switched to Gansbaai, 150 kilometres (93 miles) to the south-east. This is a white shark hotspot on account of the nearby fur seal rookeries, and researchers discovered five of its sharks were found dead, probably the work of two orcas nicknamed Port and Starboard, due to the way their dorsal fins flopped to the left or right. Since then, many more bodies have been discovered, including a 4.5-metre (15-foot) white shark that was found on the beach near the mouth of the River Nayara in May 2024. It had recently devoured a common dolphin which was in four neatly sliced pieces in its stomach. Again, the livers were gone, but Port and Starboard don't always work together. In June 2022, Starboard was spotted taking out a white shark near Seal Island, a known white shark rendezvous site on account of the Cape fur seal rookery that spawns a glut of young seals each year. Passengers on a tour boat witnessed the orca kill the shark in just two minutes, after which it appeared at the surface with the liver in its mouth. By examining bodies washed ashore, scientists have been able to work out how the orcas reach the organ and remove it with almost surgical precision.

They apply pressure on the pectoral fins and rupture the pectoral girdle to get inside. They take the liver and virtually nothing else. The surviving sharks, to state the obvious, have not been impressed. Most of Gansbaai's white sharks moved out, leaving the local shark-watching and cage diving industry with nothing to watch.

Aside from the ongoing encounters between white sharks and marine mammals, there are more 'exotic' foods to be had. The larger sharks, for example, tackle big fish, notably the giant bluefin tuna in the Mediterranean. How they catch them, though, is not entirely understood, because tuna are alert and are extremely fast swimmers. It's known that the sharks sometimes steal from the special nets of Spanish and Sicilian fishermen, the so-called almadrabas in Spain and mattanzas in Sicily. They might take large chunks out of the tuna, rendering them worthless, much to the fishermen's chagrin. One even ended up in the pen of a tuna ranch near the Coronado Islands, 13 kilometres (8 miles) from the north-west coast of Baja California, Mexico, and had to be put down. The event was quite dramatic.

The shark had detected dead fish in the bottom of the net and bit its way inside. Suddenly, it found itself in the tuna ranch enclosure surrounded by thousands of tuna – white shark heaven, but not for long. A worker shot it several times but didn't kill it, so it was thrashing around with the danger that it would demolish the entire enclosure, 45 metres (148 feet) in diameter. An Australian diver, who had been working at the farm at the time, took a powerhead, which is a spear with a shotgun cartridge at its tip, and went inside the net to try to hit the shark in the head. However, after eight blasts, the shark was still alive. The only thing for it was to haul the shark out with a crane, so the diver went inside the net again and slipped a loop over the shark's tail. When it finally succumbed to its injuries, the workers discovered that the shark was a giant about 6 metres (20 feet) long, just the sort of enormous female white shark that you might encounter at Guadalupe Island. She was probably more than fifty years old and had survived all the ravages of the ocean but came to an untimely end in a Mexican tuna ranch. For some reason the enclosure didn't have escape panels as they do at Australian tuna farms. If it had, then that shark would still be alive.

A similar event occurred in June 2002. A towboat was pulling a tuna cage, about 50 metres (165 feet) across and containing 60 tonnes (66 tons) of bluefin tuna, from Libya to Spain, when it stopped to check everything was all right. At that moment a white shark, estimated to be at least 5 metres (16 feet) long, took just five seconds to bite a hole in the net and enter the cage. The crew watched the shark but didn't try to kill it, and they didn't see it feed on the tuna. It was, perhaps, more concerned about being trapped inside the net. The boat continued on its way, and, after a couple of days, the shark left the net of its own accord by biting another hole in the net and so lived to tell the tale.

Tuna farm invasions like this by white sharks are not uncommon. Prior to 2001, there had been nine known occurrences over five years, with sharks either dead already or killed. Since then, there have been several reported. In October 2002, for instance, a large white shark was killed in a tuna cage off South Australia, and in January 2003, a 3.8-metre (12.5-foot)-long male died, possibly due to stress, in a tow cage off Boston

Island, South Australia. In June the same year, a 4.4-metre (14.4-foot) female found herself in an experimental tuna cage of the South Australia Research and Development Institute near Port Lincoln and, after a short stay, she was successfully released. Port Lincoln, in fact, has had several incursions: a 4-metre (13-foot) shark in a cage with no tuna in September 2003; two more of the same size trapped in the same cage in May 2004; and a 3-metre (10-foot) white shark in August 2004. All were released. It seems Port Lincoln's white sharks have got wise to the abundance of food within tuna cages, all except the individual that entered an empty cage – imagine the disappointment.

In the wild, swimming freely in the sea, white sharks and bluefin tuna are surprisingly evenly matched. Even though the two lines have evolved independently over the course of 420 million years, they share particular genetic traits. They both have genes that are expressed as high metabolism and the ability to keep the body warmer than the surrounding seawater, and they have a stiff body and tail that enable them to swim very fast in short bursts; in fact, these warm-bodied species can swim up to 1.6 times faster than their cold-blooded cousins. This has resulted in both being elevated to the status of super-predators, a case of convergent evolution, in which totally unrelated species evolve the same traits.

So, how – if bluefin tuna are so fast – does a white shark catch one? The shark is fast over short bursts, but the tuna could easily outpace it over a longer distance. Tuna also have eyes on either side of the head, so they have a large field of vision, making a devious attack more difficult, yet white sharks do catch large tuna, as the stomach contents of some individuals have revealed. The answer, it seems, is in the tuna's visual system. It's a diurnal animal so its vision is compromised at night; in fact, it's so bad that bluefin tuna reared in pens often smash into their walls after dark. So, night is when the somewhat slower sharks can get the upper hand and have a go at catching them using their normal sneaky hunting technique under cover of darkness.

Another large prey item is the sea turtle. There are several species and all but one – the leatherback – have a horny carapace that protects their body, so they're not an obvious white shark food. Nevertheless, they do eat sea turtles, the earliest report in modern times being from 1919 when a large 5.5-metre (18-foot)-long white shark was seen chomping on an adult loggerhead sea turtle off the coast of South Carolina. Since then, scientific papers have been rather thin on the ground, but there were two records from 1992 and 1993 of leatherback sea turtles with obvious white shark injuries that washed ashore on the California coast. The first turtle, which was 1.6 metres (5 feet) long, was found on the beach at Half Moon Bay, south of San Francisco. One of its front and one of its back flippers were bitten off close to the base. Other wounds indicated that the turtle's assailant was an adult white shark. The second was discovered on Brighton Beach to the north-west of the city. It was also 1.6 metres long and had the rear part of the carapace and a hind flipper missing. A large chunk, almost 50 centimetres (20 inches) across, taken out of the side of the carapace indicated an adult white shark about 5–5.5 metres (16–18 feet) long. In both cases, it was not clear whether the turtles had been hunted and killed or scavenged, but what it did demonstrate is that the white shark has a formidable bite. The leatherback does not have a covering of horny plates as other sea turtles, but instead has a thick shell of small, bony, tile-like plates overlaid by a thick layer of cartilage and skin. While not as impregnable as the carapaces of other sea turtles, it is still very tough, so the white shark bites that went through both the carapace on top and the plastron underneath must have had quite a force behind them. That said, the frequency of attacks on leatherback sea turtles must be relatively low. One reason might be that turtles are hard to swallow, and one individual found this out the hard way.

In April 2019, tuna fishermen in Japan described seeing a white shark swimming about with a sea turtle stuck in its mouth. Now, at the best of times, sea turtles are not easy to catch. When an attack is imminent, they tilt their body so the hard shell is facing the assailant, which means the predator will find it difficult to gain purchase. The only ways the shark is going to be successful are by adopting its usual sneaky hunting practice and surprising the turtle or by taking advantage of a sick or injured individual or one trapped in a net. However, this shark had managed to grab the turtle but now it seemed to be lodged in its formidable jaws. Sharks don't have hands so this unfortunate individual couldn't dislodge the obstruction. White sharks are of the type that must keep swimming forwards so oxygenated seawater flows across the gills. Slow that flow and the shark is in trouble, and the turtle was restricting the shark's ability to breathe. It was not in a good way, according to one of the fishermen. The next day it was found dead and wrapped in a net. Whether the turtle was the cause of death or whether being trapped in a net finished it off is unclear, but it was a very unusual event, and shark scientists are debating what actually happened to this very day. The general feeling at the moment is that the shark was regurgitating the turtle before it died and did not choke on it as was first thought. The net, as is often the case these days, killed the shark.

Surprisingly, seabirds are fair game. There are numerous sightings of white sharks appearing under seabirds and dragging them below. It's something that tiger sharks have been filmed doing in the remote north-western Hawaiian Islands, but, it seems, white sharks can be, on occasion, part of the same club. Off the South Australia coast, for example, tourists on board a Port Lincoln shark-watching boat in January 2023 witnessed a petrel fly down to examine the tuna tail bait put out to attract sharks. The bird set down on the surface and was nudged by a 4-metre (13-foot)-long white shark several times, its buoyancy enabling it to be pushed ahead of the shark's bow wave, but just as it flew towards the bait, the shark came up from behind and swallowed the bird in one gulp.

Off South Africa's south-western coast are Dyer Island and Geyser Rock, the former being the site of a penguin colony and the latter a fur seal rookery, with 'Shark Alley' between them. Here, both white sharks and fur seals prey on African penguins and several species of cormorants, the seals getting the lion's share of the birds because the sharks are more focused on the seals. At St Croix Island and Algoa Bay, 600 kilometres (370 miles) to the east, however, the sharks dominate the bird predation, especially penguins. Carcasses strewn along nearby beaches show definite signs of white shark attacks, one penguin cut completely in half, but only one penguin has ever been found in a white shark stomach. Maybe something else is going on, as all is not always as it seems. Marine biologist

Alison Kock recalls seeing a white shark sneak up on a seabird sitting on the sea's surface and then gently grab it from below. The shark swam around her boat and after a few seconds the bird popped up and flew away apparently none the worse for wear, another example, perhaps, of a white shark's curiosity.

CHAPTER 9

THE NEXT GENERATION

New Yorkers might be surprised to learn that they have a white shark nursery on their doorstep, possibly the only one in the North Atlantic. The New York–New Jersey Bight, a triangular area of sea that extends north-easterly from Cape May Inlet, New Jersey, to Montauk Point, at the eastern tip of Long Island, is where white sharks in the western North Atlantic start out in life. A tagging programme of young of the year, in the range 1.4–1.7 metres (4.6–5.6 feet) long, by scientists from the National Marine Fisheries Service of the USA, found that the young sharks remain in the area of the bight for much of the summer, from August to October, before heading south to overwinter in North and South Carolina continental shelf waters, travelling at depths down to 100 metres (330 feet). Many return to the bight in May and June the following spring. Here, the young sharks spend all of their time in the nursery area, with 90 per cent of them going no farther than 20 kilometres (12 miles) from the shoreline, while swimming parallel to the south shore of Long Island.

They stay close to the coast, not only for the food but also for their own safety. Other larger sharks could do them harm. Abandoned at birth, these babies are in uncharted territory. They will not have hunted before, or navigated these waters, but somehow they must survive, so at these sites, they learn for themselves how to navigate, evade larger predators and find and identify the right foods.

Close inshore is where they feast on the large shoals of menhaden or bunkerfish, a type of herring up to 38 centimetres (15 inches) long. They had been fished out some years ago, but a recovery programme has seen their numbers increase enormously. There are also shoals of bluefish, voracious predators that feed on small bait fish close to the surface and which migrate north and south along the east coast of the USA. Shipwrecks are also a draw, as they attract shoals of fish big and small, such as ling, cod, pollock and Atlantic spadefish that can be up to 95 centimetres (37 inches) long and go about in groups of 500-plus individuals. And, as there are many wrecks along the south shore of Long Island, including at least 140 in the so-called 'Wreck Valley' in Long Island Sound, they are popular sites for the young sharks to visit. Farther offshore, the young sharks dive for squid and bottom-dwelling flounders and rays, down to a depth of 45 metres (148 feet) or a little more. As they are less able than adult sharks to regulate their body temperature, they remain generally in waters with a temperature of around 20°C (68°F).

Along the north-east coastline, however, climate change has resulted in the sea surface temperatures having risen by 0.25°C (0.45°F) per decade between 1982 and 2016. If this starts to push the nursery eastwards and northwards, the baby sharks will end up in the same piece of sea as hunting adult white sharks, and these giants are not averse to a bit of cannibalism.

Even so, the youngsters are so inexperienced, and totally naïve about the ways of their underwater world, that they mistakenly bite people; yes, as they learn what's good to eat and what not, they make mistakes. The bites are generally not serious, as is befitting a baby shark, but even so, they can be debilitating. A 65-year-old grandmother had 9 kilograms (20 pounds) of flesh removed from her leg, just one of thirteen shark bite incidents on the south shore of Long Island during the summers of 2022 and 2023, a rise on previous years. This increase in frequency of bites is thought to be down to several factors, including warmer summers so more people in the water and the increase in the shoals of Atlantic menhaden. The sharks swim into a crowd of menhaden, slashing from side to side with mouth agape, and anybody in the way is in danger of being injured. Young white sharks are possible culprits, but there are also young sand tiger and sandbar sharks in the area too. Young white sharks, however, must make up a high proportion of those species present. It's been estimated that 200 white shark babies are delivered in the bight each year, so if a substantial number survive from one year to the next, then it has the making of a healthy population of white sharks in the north-west Atlantic.

Nobody has witnessed any of these babies being born or where or when they were conceived; in fact, only a few people have witnessed the act of procreation and none a birth. The nearest anybody has come to witnessing something close to a birth was in California, 400 metres (440 yards) off the coast of Carpinteria, to the east of Santa Barbara, when pictures from a drone picked up first a large adult shark diving towards the seafloor, about 300 metres (330 yards) from the shoreline, and then a ghostly white and slightly odd-looking baby white shark that swam up.

No longer than 1.5 metres (5 feet) with chubby, rounded fins, and swimming rather 'clumsily', it appeared to slough off a milky film from the outside of its body. Filmmaker Carlos Gauna and biologist Phillip Sternes, from the University of California, Riverside, made the discovery. The shark they watched was probably newly born, they thought, and the fluid was maybe some kind of uterine solution produced by the mother. On the other hand, it could be a sick shark with an unknown skin disease. No matter what the identity of the fluid, it was a remarkable find, the first time a newly born white shark – deformed or not – had been documented.

Until then, the smallest free-living white shark on record was caught off the Pacific coast, close to the US–Mexico international border. It was just 1.1 metres (3.5 feet) long and had features similar to late white shark embryos, although it had the same number of rows of functional teeth as an adult shark. The famous and formidable jaws and teeth tend to form quite early in a white shark embryo's development.

Both of these mini-sharks were spotted in an area suspected of being a white shark nursery. Veteran shark researcher Peter Klimley, of the University of California, Davis, identified this stretch of the Pacific coast from Santa Barbara in the north through southern California to northern Baja California, and others have noticed that young of the year have a high residency rate here throughout the year.

The young sharks prefer warm shallow waters, at a depth of less than 2 metres (6.6 feet), seeking out waters with a temperature generally between 16 and 22°C (61–72°F), and more precisely 20 to 22°C (68–72°F) if they can find it. As the sea churns and the water temperature is constantly changing, the young sharks must adjust their position in the water column to optimise their body temperature for maximum growth. According to a tagging programme by Christopher Lowe and his colleagues at California State University at Padaro Beach near Santa Barbara, they move no more than a kilometre (0.6 miles) from the shore in summer but disappear offshore in winter. While they're in their summer home, they follow a distinct diurnal pattern of behaviour.

At night, according to another tagging programme with a young 1.4-metre (4.6-foot)-long female white shark by scientists from the Pfleger Institute of Environmental Research, the sharks remain in the top 50 metres (165 feet) at the ocean surface, where they make repeated shallow, vertical excursions. From five o'clock in the morning until nine o'clock in the evening they dive deeper, down to 240 metres (790 feet), especially at dawn and dusk, gathering at the surface in the afternoon possibly to warm up. Many of these dives show additional vertical excursions, indicative of feeding at or near the seabed. During the experiment, the average sea surface temperature was about 22°C (72°F), while at depth it was closer to 9°C (48°F), showing that even young-of-the-year sharks can tolerate low temperatures when they need to. Also, during that time, about twenty-eight days before the tag was released, the young shark remained in the area of the Southern California Bight, although, as they get older, they don't stay put all the time: they start to wander.

Juveniles up to 2.5 metres (8 feet) long in the north-east Pacific hug the coast of North America, heading north in summer, returning south in winter. Significantly, the range has extended northwards, thought to be due to climate change and a warming ocean. Up until 2013, the young sharks went no farther than Santa Barbara, 34 degrees north, to the north-west of Los Angeles and not too far from their nursery area in southern California, but from 2014 they started to make it to Bodega Bay, 38.5 degrees north, 650 kilometres (400 miles) away. It's thought they find it uncomfortably warm in the south in summer, but more agreeable to the north-west of San Francisco.

One of the problems with a shift towards the north is that it might force the youngsters to adopt different diets that may not be beneficial. The juveniles of southern California that have been pushed north, for example, have taken to hunting sea otters – not a satisfying dish; in fact, they are so distasteful that they are often spat out and left to die on the shore, and sharks will have wasted energy trying to catch them.

One thing they don't do is harass people. Unlike the juvenile white sharks close to Long Island on the east coast, these west coast youngsters seem to rarely bite people. Drone surveys, as we have seen, show the sharks in the surf zone, which paddleboards and surfers frequent, and despite regularly coming within 3 metres (10 feet) of people in the water, they don't attack. People and sharks are together, as we have already learned, about 97 per cent of the time, yet most people are unaware that they are there at all. Exactly why they don't snack on people must remain a mystery for now. Perhaps they have plenty to eat and so don't need to try out something new.

Farther to the south, Sebastián Vizcaíno Bay, on the Pacific coast of Mexico's Baja California peninsula, has been identified as a white shark nursery, with the Ojo de Liebre Lagoon as the core area. About 80 per cent of records of young of the year come from the lagoon so somewhere hereabouts there must be a birthing area, although nobody has witnessed a birth. Until recently, most of the focus was on the bays on the mainland, but now scientists have been looking further afield. Located outside the bay, about 100 kilometres (60 miles) off the coast, is the dry and rocky Cedros Island, surrounded by cold but bountiful seas up to 200 metres (660 feet) deep. The large California yellowtail breeds here and is plentiful at certain times of the year, along with calico bass, a species of grouper, white sea bass, black sea bass and California halibut, all attractive targets not only for sports fishermen, but also for baby white sharks. Researchers from the Instituto Politécnico Nacional in La Paz collected records from artisanal fishermen from 2015 to 2017 which indicated the island's waters also to be a nursery for young of the year and juvenile white sharks. Fishing for white sharks is banned in Mexican waters, but many are still hauled in as bycatch. However, the sharks reported in this study were caught in gill nets and were alive when released back into the sea after data was collected. There were twelve sharks, between 1.1 and 5 metres (3.6–16 feet) long, of which six were newly born, three young of the year, two juvenile and one adult. It means outer regions, such as offshore islands, are as important as the mainland bays and lagoons to play host to white shark nurseries.

So, in summary, this is one of the sites pregnant females go to give birth. They were impregnated probably near Guadalupe Island, spent most of their pregnancy in the White Shark Café or Hawaii and went to Baja California to drop their pups. Somewhere in the area of Sebastián Vizcaíno Bay is the key to where white shark births occur in the north-east Pacific

Ocean. Now, all the scientists have to do is work out exactly where.

In the Mediterranean, the Sicilian Channel is thought to be a white shark nursery, with pregnant mothers and babies having been caught close to the tiny island of Lampedusa, famed for the large number of human migrants it receives from north Africa rather than as a home to white sharks. Across the channel, the Tunisian and Libyan coasts are possible white shark birthing places. A 5-metre (16-foot)-long pregnant female with two near-term embryos, for example, was caught at Cape Bon in north-east Tunisia, and an enormous 5.9-metre (19-foot)-long female with four embryos was hauled out of the sea in the Gulf of Gabès, just to the south-west. Attention has also turned to the northern shores of the Mediterranean, especially the Aegean Sea.

The area of sea outside the town of Altinoluk, on the northern Aegean coast of Turkey, is thought to be a white shark birthing site and nursery, with pregnant females arriving in the first week in July. Evidence for this is that in July 2008, local fishermen caught two baby white sharks here, on the north shore of the Gulf of Edremit. One was just 125 centimetres (4 feet) long, with an unhealed birthmark on its belly indicating it had been born fairly recently, and the other was 145 centimetres (57 inches) long. The sharks died and their bodies were presented to local scientists for analysis. Their DNA research revealed the young sharks were not siblings, so they must have come from separate litters, meaning at least two adult female white sharks had visited this area that year.

In July 2011, fishermen from Altinoluk caught another newborn pup in a trammel net close to shore. It's like a gill net but made up of three layers of net instead of one. Trapped as incidental catch, this time the shark was recovered and kept alive in a large round tub around which it swam continuously. A local official of the Altinoluk and Küçükkuyu Fisheries Cooperative ensured it was not harmed, and measurements were taken before the baby was released. It was a young female just 85 centimetres (34 inches) long and weighing 12 kilograms (27 pounds), although a video (which you can access on the internet) shows a shark which appears to be a bit larger. As these fishermen were not scientists, there's no knowing from which point to which on the shark's body they measured. Even so, she could not have been more than a week old and more likely four or five days old. Interestingly, the large size of her pectoral fins and tail seemed out of proportion to the rest of her body, important in keeping the little shark up in the water column although, like her Californian cousin, she swam rather clumsily. More recently, encouraged by scientists from Boğaziçi (Bosphorus) University, the members of the Altinoluk and Küçükkuyu Fisheries Cooperative are hoping to place tags on any baby white sharks that are brought in alive in the future, a plus for shark conservation.

On the other side of the world, where the Mediterranean sharks are thought have originated, white shark nurseries have been identified off the coasts of New South Wales and Victoria in Australia. The Port Stephens region in New South Wales and the Ninety Mile Beach–Corner Inlet region of Victoria are hotspots. At Port Stephens, about 200 kilometres (120 miles) north of Sydney, the young sharks, in the size range 1.8–2.6 metres (6–8.5 feet) and aged one to five years old, occupy inshore waters to a depth of about 120 metres (400 feet) along a 60-kilometre

(40-mile)-long stretch of coastline with coastal reefs offshore. They often enter the surf zone, in water less than 5 metres (16 feet) deep, where they are more visible to people. They aggregate here from early spring to midsummer and might stay in local areas for several weeks or even months where they feed on Australian salmon, mullet, snapper and mulloway or jewfish, which can grow to 2 metres (6.6 feet) long, and occasionally a stingray – perfect food for juvenile white sharks.

They stick with these kinds of food because their jaws are not strong enough to bite into larger prey, such as seals. Although their jaws and teeth may look to be just right to have a go at the bigger stuff, the reality is that they are too weak to manage the stresses. Until they are about 3 metres (10 feet) long there is insufficient mineralised cartilage in the jaws to deal with the forces involved. This might also explain why attacks on people are aborted after a single bite as the juvenile might sustain jaw injuries if it persisted with an attack.

One of the sites they visit is New South Wales's Stockton Beach, a 32-kilometre (20-mile)-long strip of sand south of Port Stephens. Not far away, at Jimmy's Beach, on the north side of Port Stephens Bay, a young woman was bitten badly by a white shark, a rare occurrence according to local shark experts. The victim had to have one of her arms sewn back on, not the kind of injury you'd expect from a juvenile. However, in 2015 a 5-metre (16-foot)-long white shark was seen cruising the area, causing beaches to be closed, so mothers must be present too, possibly indicating that a birthing area is somewhere nearby. Seal Rocks, to the north-east, is home to thousands of fur seals, so they could also be an attraction, and the coast is on the migration route of humpback whales.

From midsummer, the Stockton sharks migrate south-wards to Corner Inlet where they stay for a short time until autumn, when they move about in all directions, without any discernible pattern, eventually returning to the Stockton area in spring.

Despite many projects to find them, white shark nurseries linked to Australia's south-west population have so far evaded observers, although a blog by kayakers who were paddling near Wilsons Promontory, the dividing line between Australia's east and west white shark populations just to the south of Corner Inlet, shows a photograph of a very small white shark at the surface in the shallows, so maybe that's a possible place to look.

In New Zealand, bays on both sides of the north part of North Island, such as Hauraki Gulf to the east of Auckland and Tauranga Harbour to the south-east play host to young white sharks, known locally as ururoa. These bodies of water are thought to be pupping sites and nurseries. The youngsters generally don't go far. Like their kin elsewhere in the world, they stick close to the coast with occasional forays to the edge of the continental shelf. One little critter, tagged by Riley Elliott from the University of Auckland, though, bucked the trend.

Swaj (Jaws backwards), as the young 1.4-metre (4.6-foot)-long female was known to the researchers, found a cosy niche in a south-east corner of Doubtless Bay and stayed there. Elliott had tagged the youngster, which was less than a year old, outside Matakana Island, whose southern tip is close to Tauranga Harbour on the north-east coast's Bay of Plenty. Now, you would expect a baby white shark so young to remain in her nursery area, in this case Tauranga Harbour, but no, Swaj headed northwards, avoiding dirty nearshore waters churned up by a typhoon, by following the edge of the oceanic water. Her journey terminated at Doubtless Bay, destination of many more adventurous juveniles that start at Tauranga Harbour but travel around the tip of North Island and back again before re-entering the bay. Elliott is now trying to figure out why she just holed up in her corner and was not coming out. Local fishermen told him that there is plenty of kahawai or Australian salmon and stingrays, ideal baby white shark food, and there are octopuses and crabs. Whatever the lure, it probably involved food, Elliott believes, but after four months, she'd had enough of the same old diet and moved out.

At first, she headed northwards, skirting the Karikari Peninsula and Rangauna Bay on her way to rounding North Cape and Cape Reinga, before heading southwards along New Zealand's Ninety-Mile Beach all the way to Port Waikato, a well-known surfing spot on the west coast. She was less than a year old, but she made this extensive migration probably in search of food, although finding a comfortable water temperature might have been another factor.

Even though somewhere in the vast Tauranga Harbour is the likely birthplace for Swaj, precisely where she was born is another unknown; in fact, on the well-studied coasts of New Zealand, Australia, South Africa and North America, we still do not know exactly where and when white sharks are born or where and when they are conceived, the latter the Holy Grail in white shark research. There were, however, two momentous occasions when New Zealanders had a glimpse of that Holy Grail.

The first observation was in 1991 by seal observer Ms A.

Strachan, a temporary employee from New Zealand's Department of Conservation. She was at Nugget Point, a panoramic platform with a lighthouse on the Catlins coast named by Captain Cook because he thought the string of small, angular, rocky islets off the point resembled gold nuggets. The observer was watching the fur seal colony below the point when her attention was drawn to a commotion in the water. At first, she thought it was two animals fighting, as 'one animal appeared to be attempting to grasp the other with its great mouth, making great gouges in its side'. Eventually, the pair, which she recognised as two white sharks, became motionless, lying belly to belly, and 'turning over from time to time'. She realised instantly that she was witnessing white sharks mating, which lasted for about forty minutes, whereupon the sharks parted and went their separate ways.

The second event was in November 1997, when fisherman Dick Ledgerwood and his crewman were leaving Dunedin. When they reached Sawyer's Bay, on the way to Port Chalmers to refuel, Ledgerwood's mate spotted something white and a lot of activity in shallow water close to a sandbank. They turned their wooden boat to take a look and found two 4-metre (13-foot)-long white sharks locked together, belly to belly, and rolling slowly but continuously. Ledgerwood described it as 'rolling, rolling and rolling', each revolution taking about ten seconds. The boat drifted to within about 10 metres (33 feet) of the pair, yet the two sharks seemed oblivious to the fishermen's presence. The odd thing was that they were close to shore in just 4 metres of water. Scientists had previously surmised that white sharks must head for the deep at mating time, but here they were in the shallows – it was 'back to the drawing board' time. Clearly, it was more prudent to court and mate in an out-of-the-way place, away from seal rookeries or whale carcasses, where the male would least likely be usurped by another and the couple would be relatively safe from any large predators, such as killer whales.

What the first sighting had also confirmed is that white shark courtship is a violent affair, like that of other species of sharks whose courtship and mating have been reported, with the male biting the female around the head and on the pectoral fins. Fortunately, females have thicker skin and can take the biting without too much damage, but in locations where courtship and mating are thought to occur, female sharks are often badly scarred. Off Guadalupe Island, which could be a mating area, females show quite serious wounds and males appear to have semen in their claspers, a sure sign that mating is going on somewhere nearby, although nobody has actually witnessed it.

Fertilisation in white sharks is internal. After having grabbed the female, lain belly-to-belly and rolled around, the male inserts one of his two claspers into the female's cloaca. These claspers are the two scroll-shaped, penis-like organs that dangle from the median line of the pelvic fin on his underside. They are, in fact, derived from the inner margins of the pelvic fins. When inserted, a sharp spur folds out, serving to lock and hold the clasper in place. Semen can then be pumped into the cloaca without fear of the connection being broken.

However, to get to this stage in their life, white sharks have had to be unbelievably patient. Male sharks are at least twenty-six years old before they're ready to mate. They can be anywhere between 2.3 and 4.6 metres (7.6–15 feet) long before they get started, according to Harold Pratt, at the Narragansett Laboratory, with some of the larger sharks still showing early development of their reproductive system, even though they are of an adult size. The average male at maturity seems to be about 3.6 metres (12 feet) long. Females must wait even longer. They are generally at least thirty-three years old and 4.5–5 metres (14.8–16 feet) long before they're sexually mature, although two relatively large females, 4.72 metres (15.5 feet) and 4.96 metres (16.3 feet) long, that were caught both had small ovaries and the oviducts not well developed, which means an even longer wait before they can get going and procreate. It's understandable. Mothers must carry anywhere between two and seventeen embryos for eighteen months, and when close to term the unborn pups are each 1.1–1.5 metres (3.6–5 feet) long, so adult females have to be pretty big to accommodate such a litter.

White sharks are ovoviviparous, which means the fertilised eggs develop in the two uteri, where the developing sharks remain until birth. Work at the Okinawa Churashima Research Centre, Japan, involving a pregnant 5-metre (16-foot)-long female with six unborn pups, has shown that, during the early part of gestation, the mother secretes a lipid-rich fluid, known as 'uterine milk', from the uterine wall to sustain her young. Later in development, oophagy takes over, during which the female produces yolk-rich unfertilised egg capsules that become food for her unborn pups. The fluid becomes yellow with the yolk from broken eggs chomped on by the pups. Oophagy enables the mother to rear her unborn pups to a good, functional size before they are born.

The Japanese research revealed that each embryo has

twenty rows of 'embryonic teeth', some of which are yet to erect. They're probably used to break open the egg capsules, and teeth are often found in the stomach, not the result of intra-uterine cannibalism, but from the fact that unborn white sharks swallow their own teeth. It appears white shark embryos have functional teeth and the ability to replace them even while still in the womb. The distal end of the intestine is blocked by a plug of fine crystalline material, which is thought to prevent the embryo from defecating in the uterus. Only the upper lobe of the tail has grown, the lower lobe being very small. And the body is entirely a pale whitish colour, similar to the newborn shark in California. The mother of these pups was caught incidentally in a coastal set net off the village of Yomitan, Okinawa Prefecture, on the west side of central Okinawa Island, and, alas, mother and babies perished. The coastal waters of south-west Japan appear to be a white shark nursery area, but in the recent past they were also a place where many sharks, particularly pregnant females, were caught and taken to market.

The number of pups in a litter varies with individual mothers. The dissection of several pregnant females caught off Japan revealed one with seven embryos and another with ten. Among the record holders have been an individual reported from Australia and a 4.5-metre (15-foot)-long pregnant female caught off north-east Taiwan. Both had fourteen near-term pups inside, probably seven in each uterus. The Taiwanese shark was caught by accident and sold in a fish market to a taxidermy company for just $2,000.

Those that do make it into the world are basically small versions of their parents, albeit with smaller rounded fins and, if the California newborn shark is anything to go by, they are immensely clumsy. Their diet is very different to that of the older sharks; in fact, white sharks change their diet as they get older and bigger. Young of the year, which are less than one year old, as well as young juveniles, eat mainly midwater schooling and bottom-dwelling fish, such as flatfish, skates and rays, along with smaller sharks, squid and shoals of reef fish, although the quantity of each element of the diet can vary with location.

In New South Wales waters, for example, a study of the stomach contents of juvenile white sharks by researchers from the University of Sydney revealed precisely what their local youngsters consumed. About 32 per cent of contents was pelagic or midwater ocean-swimming fish, such as Australian salmon, whiting, mullet and eels. Close to 17.4 per cent was bottom-dwelling fish, such as stargazers, sole and flatheads, and 14.9 per cent skates and rays, such as stingrays and electric rays, although how they cope with a fierce electrical jolt from an electric ray is a bit of a mystery. They also somehow manage to successfully hunt eagle rays, yet how they are able to catch them is another unknown as these rays have a fair turn of speed. As for the rest of the food, about 5 per cent of stomach contents consisted of reef fishes, such as blue gropers, a genus of wrasse, and there was an assortment of unidentifiable fishes due to them being partly digested. Marine mammals and other sharks, as well as squid and cuttlefish, were eaten less frequently. One of the unexpected revelations was that the juveniles spend such a significant amount of time hunting fish that live mainly on or near the seabed. It's something that has been revealed elsewhere too.

In another study, this time in the USA, a tagged youngster in the Southern California Bight was found to have a diurnal dive pattern. At night, it remained in the top 50 metres (165 feet), often making repeated shallow vertical excursions. Deeper dives, down to 165 metres (540 feet), and occasionally as far as 240 metres (790 feet), took place during the day, between 05.00 and 21.00 hours, indicating that much of its daytime food consists probably of bottom-dwellers, such as flatfish and rays, species that are relatively easy to catch once they've been spotted. Unfortunately, for the shark at least, these guys are exceptionally good at camouflage, but there must be ways the sharks can detect these bottom-dwellers, or the youngsters would go hungry. One of the clues must be the electrical activity of the prey's muscles, including its heart, which never stops beating. The shark's electro-sensors in the snout will pick up this activity. For the victim, it's a dead give-away and it's probably game over.

By the time some individuals reach a length of about 2.5 metres (8 feet), closer to 3.5 metres (11.5 feet) for the slow-coaches, they will start to switch diets. While they'll still hunt down fish and squid, as they did before, in order to maintain a larger body that's kept warmer than the surrounding seawater, these young white sharks need high-energy food and plenty of it, and that means they must try out new hunting skills. To acquire the calorie-rich fats they need, they must stalk and catch highly mobile and not easily caught seals and sea lions. They need seal blubber.

This change of diet is accompanied by a change in the shape of their teeth. As juveniles, white sharks have more pointy teeth, all the better for catching slippery fish. At this stage, they could be mistaken for mako sharks, which have pointy teeth that seem to spill out of their mouths. When they become subadults, however, the teeth in the lower jaw remain kind of pointy like forks, but those in the upper jaw become the familiar triangular shape with serrated edges, the steak knives for slicing blubber.

For subadults, seals along with dolphins, bat rays, tuna and squid are the primary targets, but as they get bigger still, they tend to switch diet for a second time. The large adults, in the 4.5–6 metre (15–20 foot) category, have a predilection for whale blubber. A large female can polish off about 30 kilograms (66 pounds) of whale blubber in one sitting, which can last her for over a month without having to feed again… unless, of course, she's feeding for fourteen!

PART IV

CONFLICT

CHAPTER 10

GIANT SHARKS

One of the most controversial aspects of white shark biology must be size. Just how big do white sharks get? What individual holds the record for being the largest and how reliable is its supporting data? The tendency was for early records to be exaggerated as measuring methods were suspect. More recent reports are probably more accurate, but let's take a look at what's been recorded down the years. During this time, these giant sharks, bar a few, have all died by accident, having been bycatch to commercial fisheries, or been killed deliberately by trophy hunters and artisanal fisheries.

The first thing to say is that adult male sharks are considerably smaller than mature females, a phenomenon known as sexual dimorphism. The average length today is 3.4–4 metres (11–13 feet) for males and 4.6–5 metres (15–16 feet) for females, so most giants are females. There are, however, some white sharks of exceptional size reliably reported, taking the female maximum size to about 6.5 metres (21 feet) long and weighing close to 2,000 kilograms (4,400 lbs).

Two of the largest living white sharks are Deep Blue and Haole Girl, mature females that spend their pregnancies, not at the White Shark Café, but off the coasts of Hawaii. They haven't been measured accurately but estimates indicate a length of about 6 metres (20 feet) each. And, when they are not in the ocean around Hawaii, they can be found sometimes at Guadalupe Island where Deep Blue, at least, has starred in Discovery Channel's Shark Week. What, though, do we make of the claims of even bigger white sharks?

In 1977, a white shark measuring an estimated 9.5 metres (31 feet) was observed by anglers who were fishing for swordfish in California waters, along with the crew of their spotter plane. The fishermen, distributed among several boats, placed their vessels alongside the shark to assess how long it was and most agreed to within 30 centimetres (12 inches). Fortunately, they did not try to catch the shark, so the giant got away unscathed and unmeasured.

Also unharmed and not measured are some real monsters that lurk out there off the coasts of Australia and South Africa, if the bite marks on dead whales are anything to go by. Now, it transpires that you can work out the approximate length of a shark by the size of its bite. In May 1972, a whale killed in waters off South Australia, for example, had five chunks of blubber taken by a ravenous white shark. Each bite mark was about 60 centimetres (24 inches) across, indicating a shark with a length of 7.6–8 metres (25–26 feet). Previously, even larger bite marks had been seen on a whale in 1968, but they were not measured reliably. Even so, the individual must have been a real giant among sharks, possibly up to 8 metres long according to John E. Randall's guesstimate in a 1973 scientific paper in the journal *Science*.

Randall, of the Bernice P. Bishop Museum, Honolulu, was the scientist who worked out a formula for assessing the length of white sharks based on both jaws and tooth size. It enabled him to debunk the story of the famous Port Fairy shark from Victoria, Australia. It was caught in 1852, and its jaws retained by London's Natural History Museum. At the time, it was claimed to be largest white shark ever caught with a given length of 11.1 metres (37 feet). Randall, however, examined the jaws and teeth and applied the relevant formula and found that, when alive, it could have been no more than about 5 metres (16 feet), quite a difference!

Three more giant white shark stories come from Australia and South Africa. In his 1963 book, *Sharks and Rays of Australian Seas*, David Stead recounts a conversation with Captain J. S. Elkington about how a gigantic white pointer sidled up to his launch outside Townsville breakwater, Queensland, and lay there for half an hour, which is very odd behaviour for white sharks as they have to keep moving to breathe. The year was 1894 and Elkington describes a shark about 4 feet (1.2 metres) longer than his 35-foot (10.7-metre)-long boat, making it 39 feet (11.9 metres) long, another potential giant. And in *South African Beachcomber* (1958), Lawrence G. Green describes how a Tristan islander, who settled in Simonstown, South Africa, told him about a white shark about 43 feet (13.1 metres) long that beached itself in False Bay, Cape Province. The date is unknown.

Another tale from False Bay features a white shark that became known as the Submarine. The monster, it was said, swam from False Bay up the south-east coast and back, so children were told not to swim at three o'clock in the afternoon as that was when the Submarine was out hunting. The shark was claimed to be over 7 metres (23 feet) long, but all was not necessarily as it seemed. According to some commentators, the Submarine didn't exist. It was a story made up in the 1970s by journalists who wanted to see how gullible their readers were, but it was perpetuated down the years by fishermen and others so that people thought the Submarine was real. However, other observers are adamant that it *is* real and have spotted it or tried to catch it and it got away. So, the jury's out. There was a genuine giant in the area, though. It was the largest white shark caught off South Africa and reliably measured, at 5.9 metres (19 feet) long. The encounter was off Danger Point near Gansbaai, close to where many people have cage diving encounters with white sharks, although this shark would have scared a good few of the more sensitive souls.

The Azores archipelago, in the eastern North Atlantic, is another hotspot for giant white sharks, among the largest being one that entered the harbour of Vila Franca do Campo on the island of São Miguel in May 1978. Previously, according to sports fishing consultant and author Trevor Housby, the shark had been patrolling the harbour entrance apparently waiting for fishermen to return to port, but its entry into the harbour caused a nuisance so it was dispatched with a lance. After it had been hauled onto the quay, Housby measured the shark and found it to be 9 metres (30 feet) long. Later, however, John Randall examined a picture of the shark taken at the time and, again, he put the kibosh on the size, suggesting it was closer to 6.1 metres (20 feet) long, still a hefty beast, and that maybe Housby had measured along the curves of the body rather than in a straight line from nose to tail. The debate, though, was not over. Apparently, a tooth measuring 7.5 cm (3 inches) was examined and it indicated that the original length was probably closer to the truth. The problem now is that nobody can find the tooth! However, not to be outdone, several large white sharks, all over 6 metres (20 feet) long, have been hauled ashore. There was, for example, a 6.4-metre (21-foot)-long specimen that was harpooned close to Lajes do Pico on the island of Pico in August 1980. The Azoreans also landed another giant in February 1990. It was caught off the south coast of São Miguel, near Ponta Garça, and it measured 5.2 metres (17 feet).

How often white sharks arrive in the archipelago is unknown, but at one time, they were supposedly frequent visitors on account of the whaling in open boats for sperm whales. Whaling was conducted from 1867 to 1982, out of Lajes do Pico and also from other islands in the archipelago. The whale carcasses were cut up and rendered on the shore and sharks would gather allegedly to feed on the scraps. Reports, however, are few and far between and unreliable, so nobody knows whether this was true or not. Today, white sharks are known to occur in the vicinity of the islands, and some might arrive from an unexpected direction.

A tagging programme off the east coast of the USA found that one of its sharks had set off across the Atlantic Ocean and was tracked to within 30 kilometres (20 miles) of Flores, the westernmost island in the Azores archipelago. It's almost a mirror image of the long-distance travels of sharks like Big Blue in the north-east Pacific, and it brings up the question again of whether sharks from the north-west Atlantic ever meet up with those in the north-east.

In the north-east Atlantic, oceanic islands, such as Madeira, the Canaries and Cape Verde, have also had their share of white sharks. Strandings of great white sharks are not common globally, but one pitched up on the beach at San Cristóbal on Gran Canaria in the late 1920s, and, from 1987, there's a report of a 4-metre (13-foot)-long white shark caught off Tenerife. Jacques-Yves Cousteau and his team aboard the *Calypso* were diving at the Cape Verde Islands when they encountered what he described as 'a very large great white shark'. In his *The Silent World*, written with Frédéric Dumas and published in 1953, he describes how the shark, on spotting the divers, did something most unexpected. 'In pure fright,' he wrote, 'the monster voided a cloud of excrement and departed at an incredible speed.'

Closer to the African mainland, a specimen was caught near the village of M'Bodiène in Senegal, during the summer of 1945, and in 1957, another was seen between Dakar and the fortified island of Gorée. There have also been reports of catches from Western Sahara (where many fossil megalodon shark teeth are found) and the Atlantic coast of Morocco, where a large female white shark, more than 4 metres (13 feet) long, was caught and displayed on the quay at Kenitra in 2016. It was caught by a tuna fishing boat.

However, reliably measured giants are not common. There's nearly always something wrong with the data to bring the measured length into doubt, but there are a few specimens which really are trustworthy. The largest of these is a white shark that was brought into Sète, a coastal town to the south-west of Montpellier in southern France and bordering the Gulf of Lyon. The date was 13 October 1956, and the time early morning when the shark was caught in a tuna drift net set 4.8 kilometres (3 miles) offshore from Maguelone by the fishing vessel *Rosina-Raphaël*. The specimen was a female that was

5.89 metres (19.3 feet) long when freshly caught, and when cut open she had two partly digested 1.8-metre (6-foot)-long dolphins in her stomach. Overseeing the measuring and dissection were representatives of the Marine Biological Station in Sète and the University of Neuchâtel, Switzerland, so it was all above board. And we can be sure of the data because a model was made of her, the largest mould of a white shark constructed from casts of an original specimen by taxidermist Eugène Küttel, meaning that her complete morphometrics – 'the quantitative analysis of form' – are recorded, the only white shark to have had this. Her fins and teeth have been incorporated into the model, which now resides in the Cantonal Museum of Zoology in Lausanne, Switzerland, but it's not quite exact. During the mould-making process, the shark shrank by 6 centimetres (2.4 inches).

It wasn't the only white shark to have been caught off Sète. There have been three others reported and included in a paper in *Marine Life* by Alessandro de Maddalena, of the University of Milano-Bicocca. They are a 4-metre (13-foot)-long shark in 1875, a 2.4-metre (8-foot) specimen the following year and, more recently, another giant female about 5.9 metres (19 feet) long in January 1991. She was acquired by a wholesale fishmonger, transported to the Rungis International Market in Paris and bought by a supermarket in Montargis to the south, a total overland distance of 880 kilometres (547 miles), making her possibly the most widely travelled white shark… on land!

The Mediterranean, as de Maddalena would attest, has seen an unexpectedly large number of giant white sharks, some turning out to be not so giant, but others joining the ranks of the mega-sharks. Here are a few.

In February 1839, a large white shark was hauled from the sea near Civitanova Marche, equidistant between Pescara and Rimini on Italy's Adriatic coast. It was not measured but, based on the size of its largest vertebra, it is thought the animal was close to 6 metres (20 feet) long. In 1886, a white shark estimated to be a staggering 8–9 metres (26–30 feet) long was caught near Piombino on Italy's Tyrrhenian coast. Little else is known about it, but in the same area, in the Bay of Baratti, close to the island of Elba, on 2 February 1989, there occurred a devastating shark bite incident. Luciano Costanzo was spearfishing in scuba gear and was returning to his dive boat, where his son and a friend were waiting. They looked on in horror as Luciano surfaced rapidly and shouted 'Shark' before making for the boat. He didn't make it. The shark grabbed him around the waist and pulled him below, surfaced three times and then disappeared. 'I clearly saw my father in the mouth of the shark,' his son said later, 'and the sea was red with my father's blood.' Three days later a dive team searching with an underwater camera came across flippers, a weight belt and a pair of scuba tanks with shark bite marks but no sign of poor Luciano. It was the first shark attack in the Mediterranean for thirty years and, understandably, it drew in the world's press. Depending on which newspaper you read the shark was variously described as '20 feet (6 metres) long', 'at least 22 feet (6.7 metres)' and '25 feet (7.6 metres)'. If truth be told, nobody knows, but it must have been pretty hefty.

On the night of 12 August 1938, a large white shark was caught in a tuna trap – a tonnara – belonging to the Rid brothers of Enfola on the north coast of Elba, clearly a white shark hotspot. The shark was killed with harpoons and rifle fire and towed to land. Here, the stomach was cut open and out spilled two dolphins. It was not measured but cut into pieces which were shipped to a market in Florence. The only evidence remaining is a couple of black-and-white photographs showing the shark with people standing behind it. The estimated length, calculated by de Maddalena, depends on the height of a girl standing next to the shark's snout. On the one hand, if she was 1.70 metres (5 feet 7 inches) tall, the shark was probably 5.97 metres (19.59 feet) long, but on the other hand if she was 1.75 metres (5 feet 9 inches) tall the shark would have been 6.13 metres (20.11 feet) long. The girl is relatively tall compared to the other people in the picture, so the greater length might well be the correct one.

On 19 June 1961, fisherman Domenico Sorrenti and his crew of four were off Ganzirri, at the northern end of the Strait of Messina, a lively part of the Mediterranean, where unusual hydrological conditions result in whirlpools, origin of the mythical Charybdis. Sorrenti's encounter, however, was far from mythical. He set out that morning and round about noon he spotted a large dorsal fin about 150 metres (164 yards) away. He caught up with the shark, a large female white shark, which he harpooned. At first, it didn't seem to put up much of a fight and dived to the bottom. The fight, however, lasted about an hour.

The other fishermen came to help but the shark was so big it was impossible to get it into the boat, and so Sorrenti towed the animal to shore and hauled it out on the beach. He estimated it to be approximately 6.4 metres (21 feet) long, and when he cut open the stomach out fell a large, recently eaten dolphin sliced into two pieces. Sorrenti thought that the full stomach might have been the reason the fish did not fight as strongly as he would have expected. No actual measurements were taken but when de Maddalena examined the blurred photograph of the shark on the beach, he thought that Sorrenti's original estimate was close to being right. In 1965, another giant was caught here, estimated to be close to 6.2 metres (20.3 feet) long, although revisions down to 5.6 metres (18.4 feet) have been suggested because it's thought the measuring tape followed the curves of the body rather than the straight-line distance from nose to tail.

Previously, a white shark in the Strait of Messina was harpooned in 1913. The fisherman's son, who still has the jaws in his possession, recalled that it was 6.5 metres (21 feet) long and it had two bluefin tuna in its stomach, each weighing about 100 kilograms (220 pounds) and both in two pieces.

On 2 October 1963, fishermen were hauling their nets aboard a fishing boat from the Delamaris fish-processing company when it was approached by a large white shark – bad mistake! Twenty-three rifle shots struck its body, and the shark was dead in the water. They were off the Adriatic coast of what was then Yugoslavia, in the Gulf of Piran, close to the coastal settlement of Savudrija, although the fish was hauled out on the quay at Izola, a little way along the coast to the north-east. Cutting open the stomach revealed the shark had just eaten a 200-kilogram (440-pound) dolphin. No measurements were taken but it was photographed, and the picture ended up in a scientific paper by Dejan Bošnjak and Lovrenc Lipej in *Proteus*, but Professor Lipej wanted more detail, so he found a fisherman who was involved in the capture, and he estimated the shark to be about 6 metres (20 feet) long. To check if this was close to the truth, Lipej had a spark of genius. He noticed that the shark was resting on pavement blocks on the quay whose edges were very clear. He went to the quay and measured them. Returning to the photographs, Lipej found that the shark lay on ten blocks from the snout to the tail. In the photograph, the body is slightly curved, so he had to compensate for this, but he came up with a length of about 5.5 metres (18 feet), considerably less than the fisherman's estimate. The technique was simple but clever, and it turned out that the shark was not among the major giants.

Switching to the western Mediterranean, some pretty big white sharks have come close to the Balearic Islands. In February 1969, a female was caught near Mallorca that was purported to be an astounding 8 metres (26 feet) long, but again examination of photographs points to a length nearer 6.4 metres (21 feet), still a formidable shark. And, in 1976, another one appeared, also estimated to be 6.4 metres long. In fact, it might shock European holidaymakers that the Balearic Islands appear to regularly attract white sharks, and in a paper in *Environmental Biology of Fishes*, Gabriel Morey and colleagues documented all the white sharks present in the Balearics from fisheries catch records, dating from the 1920s to the 1970s. They came up with twenty-seven reports of white sharks with lengths variously between 3.24 and 8 metres (10.6–26 feet). Nineteen claimed to be 5 metres (16 feet) or more, with three at 7 metres (23 feet) and the 1969 shark at 8 metres (26 feet) but later revised. All had been caught in bluefin tuna traps (almadrabas), a method of fishing not used in the Balearics since the late 1970s. White sharks are present all year round, with a concentration in winter off the north-eastern coast of Mallorca, where the continental shelf is relatively wide.

May 1974 saw the focus switch to Formica, an islet at the western end of Sicily, to the north of Marsala, where another tonnara bagged another large white shark. Boss of the trap, known as the 'rais' in Sicily, was Michele Grimaudo and together with the inspector of tuna trap nets, Nitto Minneo, he examined the shark. Minneo dived down and found that the shark was dead, its head caught in the net. He extracted the teeth and distributed them to the crew. When the stomach was cut open a motley bunch of articles fell out including plastic bottles and shopping bags and – a goat! The shark was measured several times under the watchful eye of the rais, and Minneo reported its length as 6.2–6.4 metres (20.3–21 feet). When de Maddalena and his colleagues came to look at a photograph, they ascertained that the shark was closer to 5.94 metres (19.5 feet) long – nevertheless, a big shark.

In September 1979, a 6.2-metre (20.3-foot)-long white was caught 10 kilometres (6 miles) off the coast of Gallipoli (Puglia), in the same place where one was caught six or seven years previously. It was trapped in a net set at about 30 metres (100 feet) and was alive when the fisherman Pompeo Alessandrelli discovered it. Harpoons were used to dispatch it, and the body was strapped alongside the fishing boat and taken into the harbour in Gallipoli. The shark was so big, the crane of a cargo ship in the port had to be used to lift it onto the quay. Before it was cut up, the shark – a male – was examined and measured by Francesco and Antonio Piccinno on behalf of the University of Lecce. When they arrived, they found that the shark had caused quite a stir on the quayside, with hundreds of people having to stay behind makeshift barricades. They also had a shock when opening up the shark's stomach. Inside were a pair of shoes that hadn't been digested, the remains of a tragic meal.

If any shark has drummed up a deal of controversy, it's the Filfla specimen, originally quoted at 7.4 metres (24 feet) long, which would have made it one of the largest white sharks in the world, but the length and weight have been questioned mercilessly. The shark was caught to the south of Malta off the tiny island of Filfla on the morning of 17 April 1987, by fisherman Alfredo Cutajar. He, along with several other fishermen, was south of Filfla; in fact, the shark was caught originally by a line belonging to a Mr D'Amato but his line snapped and became entangled with those of Cutajar, who landed the shark. At first, it was taken to Wied-iż-Żurrieq, a diving centre on the south coast of Malta, but access is difficult, and the harbour winch was not strong enough, so it was dragged around to Marsaxlokk, a traditional fishing village in the south-east corner of the island. Here, it was sliced open and in its stomach was a whole blue shark about 2.2 metres (7 feet) long, a 2.5-metre (8-foot)-long dolphin of indeterminate species bitten in half, a loggerhead turtle with a 60-centimetre (24-inch) carapace and a plastic bag containing garbage.

That evening the enormous shark was kept in a large garage overnight and taken to Valletta fish market in the morning. However, the jaws were preserved by John Abela and exhibited at the Museum of History and Culture on the nearby island of Gozo, off Malta's north-west coast. Abela also measured the shark, and he seemed to know what to do: to measure not on the curves but in a straight line. He came up with a length of 7.14 metres (23.4 feet), although at various times later he quoted several different lengths, calling into doubt his original calculation, not least through his mention of comparing it to the pickup truck transporting it, indicating, perhaps, that measurements hadn't been taken after all.

Several good photographs exist of the shark on the quay, and Mediterranean shark aficionado Ian Fergusson published that it was in the region of 5.2–5.5 metres (17–18 feet) long. While making a BBC documentary about white sharks in the Med, the producers called upon British forensic investigators to examine the photographs and give their verdict – 5.5 metres (18 feet) long, they said. Under pressure from reporters on the same programme, Abela said, 'I may have made a mistake taking the measurement.' But the story didn't end there.

Alessandro de Maddalena and his team had access to unpublished photographs and picking three that were suitable for analysis he calculated that the total length of the shark was 6.81 metres (22.3 feet) in photo one, 6.78 metres (22.2 feet) in photo two and 6.68 metres (21.9 feet) in photo three. The consistency of the three figures seems to indicate that Abela's original estimate of 7.14 metres was not far off the mark.

One part of the Mediterranean where you might not expect to come across gigantic white sharks is the Bosphorus area of Turkey, the narrow strait that joins the Marmara and Black Seas and is the continental boundary between Europe and Asia, but you'd be wrong. There have been some stonkers caught around here. Since 1916, four white sharks estimated to be 7 metres (23 feet) long have been reported, and there are two specimens that are claimed to have been 8 metres (26 feet). The first was caught at some time before 1926, and it had two large bluefin tuna and a large dolphin in its stomach. The second was caught on 18 December 1958. It was off the coast of Ahırkapı, where the main shipping channel from the Sea of Marmara to the Bosphorus and the Black Sea begins. This second shark was measured, but scientists wonder again if the measuring tape followed the curve instead of a straight line.

All this means, of course, that the Mediterranean region as a whole appears to have been blessed, if that's the right word, with a surfeit of exceptionally large white sharks. The reason is that several parts of the Mediterranean Sea are birthing sites for pregnant white sharks, and pregnant sharks are the largest sharks. Many of the sharks mentioned above were probably going to give birth, had recently given birth or were cruising for an assignation, but the Mediterranean doesn't have a monopoly on these giants. Elsewhere in the world, there have been even more real colossi.

During the 1930s, two large white sharks on their summer holidays visited New Brunswick and the Bay of Fundy on Canada's Atlantic coast. The first arrived mid-June 1930 at White Head Island where it got itself stuck in a herring weir, an ancient and ingenious way of trapping young herring. The weir works by allowing the fish to enter a 'neck' which leads to a cage-like area made of poles and brush. Once in, they find it difficult to get out. The Passamaquoddy people constructed the weirs many years ago, and they are still in use today. Should a white shark become trapped inside, then it would not only destroy the flimsy structure but also become entangled in it and, unable to move forwards, it would drown, and this is precisely what happened. This shark, however, was of an exceptional size, at least according to the fisherman who found his weir in pieces and the dead shark wrapped up inside. After extracting it and hauling it to the quayside, it was measured, but how accurately this was, is hard to say. The figure quoted is 11.3 metres (37 feet), which would make it a record holder, but, as the method of measuring was uncertain, scientists cannot verify the claim.

A second visit to New Brunswick by a memorable white shark was about 85 kilometres (53 miles) to the north-west at Campobello Island. It was 22 November 1932, when most white sharks would have headed south for the winter, but this shark stayed for a bit longer. It turned out to be a bad decision because, like its predecessor, it got itself caught in another herring weir in the Harbour de Lute, a known good fishing spot for sea bass, an attraction maybe for a white shark. When measured, it was found to be 7.9 metres (26 feet) long – an exceptional size for a white shark. When opened up, the liver was reported to be huge, 'exuding a great quantity of oil which was sold to an oil factory', enough to fill seven 50-gallon (190-litre) drums. The remarkable thing about this story is that it was not the first time that this shark had been encountered.

In July of the same year, fisherman Wilson Munroe and his young son were attending their lines 16 kilometres (10 miles) north-west of Digby Gut on the eastern side of the Bay of Fundy when a large shark, longer than their boat, began to circle them. Without warning, it slammed into the 7.6-metre (25-foot)-long dory, causing the starboard side to rise and the port side to sink and water to slop over the port gunwale. The shark bumped the boat several times, working its way towards the stern, before it disappeared altogether. When the fishermen returned to harbour, they hauled out the boat to check for damage and found the blades of their three-blade propellor bent and teeth embedded in the keel and elsewhere. These teeth and those from the fish weir shark were thought to be from one and the same individual.

Way to the north, on 4 August 1983, Prince Edward Island played unwitting host to a giant female white shark. She was trapped in a cod net off Alberton in the north-east of the island, facing the Gulf of St Lawrence. She was hauled out and onto the quay by a crane and local people came to stare at the giant. In photographs, the shark dwarfed anybody standing beside her. Scientific papers variously gave her length as anything between 5.2 and 6.1 metres (17–20 feet), which seems to indicate that nobody actually measured the creature. However, before anybody could do so, fisheries officers appeared, took away the corpse and buried it!

In the western part of the Indian Ocean, Malindi, at the mouth of the Sabaki River on the Kenyan coast, is set among a string of tropical white-sand beaches and coral reefs and not a place, perhaps, that you'd associate with white sharks. One turned up on 16 June 1996, though, and it was a big one, a pregnant female. The shark was trapped in the fishing gear of an artisanal net fishery, set at night at a depth of 35 metres (115 feet) and about 10 kilometres (6 miles) offshore. The fishermen brought her ashore but cut her into pieces before she could be photographed and measured. The parts were weighed, although a lot of body fluids were lost and the six to eight near-full-term embryos, each about 1.1 metres (3.6 feet) long, were excluded from the weigh-in. Several more embryos were aborted during the capture, so the litter was thought to be up to seventeen youngsters, a world record. Nevertheless, scientists were able to work out that her total length was about

6.4 metres (21 feet), another giant. And, if her embryos were about to be born might there be a white shark birthing area nearby? Watch this space. One of the locals mentioned that several white sharks had been caught in the area in the recent past, but this pregnant female was by far the largest.

Sabah (Borneo) is another location that you wouldn't associate with a white shark, but between 12 April and 5 May 1981, commercial divers encountered and photographed a white shark where they were laying an underwater pipeline from a barge close to St Joseph's offshore oil field in the South China Sea. The shark appeared for a couple of days during that period, and it represents the first reported occurrence of a white shark in Borneo waters, but what a debut. It's a corker, and this giant didn't end up dead. The divers estimated that it was 6.7 metres (22 feet) long by comparing its length to a standard 12.2-metre (40-foot)-long section of pipe. At the time, it was also, according to Clifford Duffy, from the University of Auckland, who compiled the report in *Marine Biodiversity Records*, the southernmost confirmed record from the north-west Pacific Ocean.

On 15 March 2008, an immense white shark, if the pictures are anything to go by, was caught off Sandun, China. Chinese sources claim the shark was a staggering 10 metres (33 feet) long. However, when it came to weigh the beast, a weight of 5,000 pounds (2,270 kilograms) was recorded. This would be more in keeping with a shark 6.1 metres (20 feet) long, still a large animal but not a record breaker.

However, for the largest ever white shark accurately measured and verified we have to go to Australia and a female that was caught off Ledge Point, Western Australia, on 22 March 1987.

Gordon Hubbell, of Jaws International, Key Biscayne, Florida, had the jaws (now donated to the Florida Museum of Natural History along with his vast collection of shark jaws and teeth) and he verified that the shark was 5.94 metres (19.5 feet) long.

In April the same year, the 1st (April Fool's Day) to be precise, commercial fisherman Peter Riseley from Robe, a fishing port on South Australia's Limestone Coast, was about 75 kilometres (47 miles) south-south-east of Cape Hart on Kangaroo Island, to the south of Adelaide, when an ordinary working day turned into something quite unexpected. It was getting towards dusk when he and his two crewmen were retrieving the bottom-set gill nets close to the seabed at a depth of about 60 metres (200 feet). One of them yelled that they'd snagged a giant skate, but as the unusual catch reached the surface, they realised that a very large, female white pointer had become entangled in the gear and had died.

It was the second Riseley had caught in thirteen years. This one, though, was huge; in fact, too big to bring aboard. At first, they brought it to the stern of the boat and struggled to remove the net. They then tried to use the boom winch to haul the creature onto the deck, but it was having none of it. The hydraulic winch was supposed to be able to lift in excess of 2,000 kilograms (4,409 pounds), but it strained so much that the hoses came loose. In order to lighten the load, they hacked off a 3-metre (10-foot) section of tail and the dorsal fin, but the winch still wouldn't budge. Then they decided to sever the head from the body, but that was easier said than done. The three fishermen took it in turns to hack through a head that was more than a metre (3.3 feet) across, and after four-and-a-half hours it was completed. Using the partially operating winch, they managed to get it onto the deck. However, with the shark in pieces and the bulk of it still in the water, it was impossible to get any reliable information about its size, but Riseley did note that it was strung up between the fore and aft stanchions on his boat, which made it a total length of roughly 6.9 metres (22.6 feet). However, the shark was sagging in the middle so he thought that might be an underestimate and it was probably closer to 7 metres (23 feet).

One useful thing Riseley did was to retain the head and remove the jaws, so scientists had something to work with. Michael Cappo, from the Australian Institute of Marine Science, was one researcher who took an interest. He measured the teeth and jaws and used the relevant formulae to assess the shark's length when alive. The first thing that struck him was that the teeth were remarkably small for the width of the jaws – precisely 0.942 metres (3.09 feet) across. As a consequence, extrapolations using tooth size gave a smaller body length than those using the jaw size – 5.8 metres (19 feet) compared to a record-breaking 6.8 metres (22.3 feet), this latter figure closer to Riseley's estimation. Game fishermen who have seen the photographs and the jaws believe that the shark's total length is probably more like 7 metres.

The jaws were taken to Brian Tonkin of Fishton Agencies in Adelaide. He cleaned and set them for mounting. Ken Jury, editor of South Australian Fisheries' *Safish* magazine, asked Peter Riseley what he was planning to do with them.

'I'll hang them behind the toilet door,' he said, 'and scare the redbacks [spiders] away!'

In actual fact, Riseley exhibited them at his home town of Robe and collected money for the South Australian Country

Fire Service, which serves country communities and is manned by dedicated volunteers.

Finally, our trundle through the growing list of giant white sharks is not complete without reference to a story from west Africa, where a giant among giants was caught off the coast of Dakar, Senegal, in the eastern Atlantic. Waiting at quay for it to be landed was Dr Juan Antonio Moreno, a member of the European Elasmobranch Working Group and a distinguished ichthyologist, but for some inexplicable reason, he was not given permission to measure or even photograph it. So instead, he used his feet to measure surreptitiously from one end of the shark to the other, and was able to come up with a rough length of over 8 metres (26 feet), making it one of the largest sharks to have been caught in recent years. Due to the unusual nature of the data acquisition, Moreno never published a paper on the subject, and now no trace of the shark is left. As he understands it, the jaws were sold to an American customer for $1,000 and are now lost to the scientific world. Ah well.

CHAPTER 11

TROPHY HUNTERS

Until quite recently, it was always open season for white sharks the world over, and many hundreds have been killed to satisfy the vanity of sports fishermen – the trophy hunters – as they do battle with one of the ocean's most awesome apex predators. Negative though the prospect was at the time, there were some positives that came out of it. Shark bodies brought ashore, for instance, provided scientists with valuable data, which is rarely available today because, in many parts of the world, white sharks are protected and so catching and killing them is illegal. In the 1970s, this wasn't the case.

In July 1975, for example, Captain Paul Sundberg tried to catch one of the giants. He was aboard his charter boat with a party of sports fishermen, and they were about 34 kilometres (21 miles) south-south-east of Montauk Point, the eastern tip of Long Island, New York, when a white shark estimated at first to be 8.2 metres (27 feet) long passed underneath the vessel. Sundberg recalled the encounter to veteran animal life chronicler Gerald L. Wood, who compiled the entries for *The Guinness Book of Animal Facts and Feats*, a meticulously researched volume which includes some of the more contentious white shark records. When the shark approached the boat again, Sundberg and his mate stabbed the animal with two harpoons and, after a two-hour struggle, the shark broke free. Returning to harbour, Sundberg discovered that the monster shark had left a 76-centimetre (30-inch)-wide bite mark on the boat's stern plus a fragment of tooth. A blue shark which had been tied to the stern was bitten in half. Working out the approximate length of a shark based on the size of its bite mark revealed Sundberg's shark must have been close to 9.1 metres (30 feet) long.

Montauk, meanwhile, continued to present us with more giants... or was it the same one? On 23 June 1978, charter captain John Sweetman was aboard his boat *Ebb Tide*, with his son, his mate Jimmy and six paying sports fishermen. They were about 23 kilometres (14 miles) out of Montauk in about 55 metres (180 feet) depth of water, when they spotted the dorsal fin of a large shark. It was roughly 9 metres (29.5 feet) long and weighed in the region of 2,300 kilograms (5,000 pounds), Sweetman thought, and coming up behind it, Jimmy stuck it with a harpoon. The shark promptly shot off close to the surface, dragging 7 metres (23 feet) of steel cable, 100 metres (328 feet) of thick nylon line and two large kegs skipping over the water. In a scene reminiscent of *Jaws*, the shark sounded, and the kegs disappeared. When the shark surfaced, the line was recovered and tied to a stern cleat, and then it began to tow the 12-metre (40-foot), 16-tonne/ton boat backwards, that is, the way of most water resistance, at a speed of about 2 knots (3.5 km/h) and occasionally a little faster. Sweetman called Sundberg, who was a pal of his, for help; coincidentally he was no more than 5 kilometres (3 miles) from the position at which Sundberg had struggled with a shark of similar size in 1975. As Sundberg got closer, ready to stab the shark with another harpoon, he wondered whether it could be the same shark. Sweetman's mate Mike spotted the shark just 1.5 metres (5 feet) below the surface and shouted 'Oh my God, oh my God, oh my God!' according to newspaper reports as, by now, the press had descended on the two boats, including a helicopter news crew that got so close it blew Mike's hat off. However, the noise of the helicopter caused the shark to dive every time Mike was about to throw his harpoon. In the growing darkness, about fourteen hours after it was first harpooned, the shark made one last dive and the line parted, causing Sweetman's boat to stop in its tracks. Sundberg is reported to have had the same feeling as before: 'Your heart goes out,' he said, 'your knees go weak, you feel like hell; there he goes again,' although *he* is more likely to be a *she* at that size, and Sundberg agreed with Sweetman's estimate of the shark's length.

There was dead silence for a moment, and then Sweetman said quietly, 'Well, the fish won,' and the disappointed captain headed for home, only home had become a chaotic round of press interviews, photographs and phone calls from as far away as Hong Kong and Australia. Sweetman had become a reluctant celebrity. However, not everyone was happy about this kind of attention. The manager of the largest local resort hotel, echoing the sentiments of the mayor in *Jaws*, was afraid that the publicity would stop holidaymakers from coming to the white sand beaches at the eastern end of Long Island, and he was even obliged to close his beach when a fin was seen above the sea's surface.

This all took place about the same time as *Jaws II* was released, and so some folk thought the Sweetman–Sundberg circus was simply a hoax, a stunt to publicise the film. The two protagonists and some of the shark scientists who often collaborated with them knew otherwise. However, it didn't stop the theatrical producer David Merrick offering his congratulations to David Brown, one of the producers of *Jaws* and *Jaws II*, when they met in the street in New York. Brown thought Merrick was praising one of his films, but actually he was referring to the successful publicity. Brown, of course, denied it and still tells the tale at cocktail parties. He told Merrick, according to a report in the *Honolulu Star*, that 'neither he nor Universal Pictures had anything to do with the shark off Montauk, that the logistics of such a stunt would be mind boggling, and, besides, he said, he has a place on eastern Long Island and is not about to bring down property values there!'

Anyway, two days after Sweetman's little adventure, another skipper harpooned a shark which was thought at the time to be the same one. It too broke the lines when it sounded and escaped. If it was the same shark, it has clearly led a charmed life and, as white sharks live so long, it could still be out there beyond what Montauk locals call 'The End'.

The hamlet of Montauk is a globally recognised fishing centre. Fishermen from here have broken more saltwater fishing records, it is claimed, than anywhere else in the world, and one name more than any other has been at the centre of that success – Captain Frank Mundus. In 1951, he began fishing charters targeting bluefish, a strong and lively game fish that's typically 20–60 centimetres (8–24 inches) long but has been known to reach 1.2 metres (4 feet) long and weigh 14 kilograms (31 pounds). The only problem for Mundus was that they were not plentiful in Montauk waters. Sharks, on the other hand, were. He switched to big shark fishing, what he described as 'monster fishing', and he was good at it.

At first, he killed whales to make the evil-smelling chum that was ladled into the water to attract the sharks, and he also used the meat as bait. Naturally, this brought criticism even in those 'anything goes' days, but Mundus shrugged it off. He was not only a fisherman but a showman. He'd bring in the dead sharks and string them up on the quayside so everyone could see, and his arrival in port with several very large white sharks turned shark fishing into a popular spectator sport. He said himself that 'the charter business was 90 per cent show and 10 per cent go'.

He was, as most friends and colleagues would probably agree, an eccentric. He wore an Australian slouch hat, a shark tooth necklace and a hoop earring, and he even painted his big toenails, one red and the other green, for port and starboard. It is probably no coincidence that Peter Benchley was frequently on shark fishing expeditions aboard Mundus's charter boat *Cricket II*. The boat's name was inspired by Mundus's view that his profile resembled Jiminy Cricket, a fictional character created by Italian writer Carlo Collodi for his children's book, published in 1883, with the title *The Adventures of Pinocchio* and featured in the Disney animated film *Pinocchio*, made in 1940. Benchley, of course, went on to write his own famous book and it's generally believed that the character Quint is based on Mundus.

The highlight of Mundus's shark fishing career began one Sunday afternoon. The date was 10 August 1986, when Mundus took out some businessmen to go and catch some small sharks. While out on the ocean, they came across a dead whale along with a shiver of sharks feasting on the energy-rich blubber. Mundus had an inkling that a big one might turn up, so he wanted to stay with the floating carcass overnight. His charter people had to get back to their offices on Monday, so another boat arranged to take them back to port. The stakeout paid off. On Monday afternoon, a very large white shark came to the carcass. Many of the smaller sharks gave way to the bigger specimen, so Mundus could clearly see his target. However, his age was catching up with him and he was unsure that he could play a shark of this size for however long it took to land it. He called up a friend, Donnie Braddick, who was out on his boat *Fish On* and fishing for tuna. He would have the stamina to battle this shark, Mundus thought. Braddick couldn't resist the invitation and came over, lashing his boat alongside the *Cricket II*. Baited with butterfish and whiting, Mundus played out his line, while Braddick took the helm. It took the best part of an hour to hook the fish, but they did it. It was time to swap seats. Braddick took the fighting chair, Mundus the helm.

By this time, the news had got out that Mundus and Braddick were battling a mega-shark, and by eleven o'clock that night a thousand people had come to the quayside in Montauk to witness its arrival. Car headlights lit up the harbour and eventually the *Cricket II* appeared, motoring very slowly and with quite a list as the shark strapped to its side was so heavy; in fact, it was so big that a forklift truck standing by to haul it up was not up to the task. A crane from a construction site had to be purloined, and together with the help of a heavy duty cargo net the giant shark was lifted onto the scales. Captain

Carl Darenberg of the Montauk Marine Basin officiated and announced the weight – 1,555 kilograms (3,427 pounds), a new world record for a white shark taken on a rod and reel. Unfortunately, a local reporter had interviewed Mundus after the epic battle and had written about the swapping of seats. This was a big no-no. The International Game Fish Association declared that, because neither Mundus nor Braddick had played the shark from the moment the bait was set to its final demise, the weigh-in was null and void. It was a huge disappointment, but today, a fibreglass copy of the record breaker that never was can be seen strung up on the Lake Montauk waterfront just as the real shark would have been in Mundus's time.

The actual world record holder for a white shark taken on rod and reel is the Australian big game fisherman Alf Dean. Dean was a very different character from Mundus: quiet, humble and only a part-time fisherman; after all, he had a citrus and grape plantation to run, and fishing had to fit in between one harvest and the next. Among other harbours, Dean often fished out of Port Lincoln, South Australia, a fishing centre that the American author and lifelong fishing buff Zane Grey described as 'a game fishing port rivalled by few in the world'. Grey was there before World War II, but after hostilities had ceased Port Lincoln took on a new lease of life. It was then that Dean began chasing sharks and records.

Dean's first white pointer was not a particularly auspicious occasion. He hooked a shark, but it promptly broke his rod, and the line had to be cut to save the rest of it. A crewman tied the handle of a broom to the surviving piece of the rod and Dean hooked another fish, a modest white pointer weighing 394 kilograms (868 pounds). The other fishermen pulled his leg mercilessly: 'Not bad, Alf, for landing him with a broom-stick!' Not to be outdone, Dean then made his own special rod.

However, records were won and lost regularly by others. Australian sheep farmer E. H. V. Riggs set the bar before the end of the war, in 1941, with a white shark that weighed 792 kilograms (1,747 pounds). He'd caught it before, but it got away. He recognised it the second time because it still had his hook in its mouth. His record was beaten by well-known Port Lincoln big game fisherman Jim Veitch with a white shark weighing 795 kilograms (1,752 pounds).

In 1950, a lot of interest was shown in the industry when the governor of South Australia, Sir Willoughby Norrie, mounted a three-day expedition to the Joseph Banks group of islands in Spencer Gulf, which includes Dangerous Reef, where you'll recall the wildlife sequences for *Jaws* were filmed. He hooked, played and landed a huge shark weighing 1,009 kilograms (2,225 pounds) and the press was watching. It was a huge boost for Port Lincoln and, despite Norrie's illustrious career in the diplomatic service, he forever claimed that his greatest achievement was catching that shark. Alf Dean came in not long after Norrie with a modest 390 kilograms (860 pounds).

By this time, Dean had got the bug and really wanted to go for the big sharks, the record breakers, and the place where one in particular occurred was Streaky Bay, on the western side of the Eyre Peninsula. The shark was nicknamed Barnacle Bill, and it was supposedly an enormous white pointer, the kind that breaks records. The bay itself is called Blanche Port, and Dean asked for his boat to be anchored just outside, near the anchor light at the entrance of the bay. No lines were set that night, but whale oil was dripped into the sea to keep the sharks interested. By two o'clock in the morning the boat was being bumped, and at dawn the crew could see two sharks, a big one and a smaller one. It was time to engage with the big one, and baited lines were deployed. The larger shark took the bait, and the fight was on, although it lasted no more than ninety minutes. Back at the quay, the shark was found to weigh 1,058 kilograms (2,333 pounds), another record catch, and measured 4.95 metres (16 feet) long. However, Barnacle Bill eluded Dean. He was beginning to wonder if it existed at all, but so many people had spotted it that it must have been real. In the meantime, Dean took another big one with a weight of 1,066 kilograms (2,352 pounds), which meant he'd broken his own record again. The newspapers were full of it, and even Hounsells of Bridport, the UK twine and rope manufacturer, took out an advert which declared that Dean had used their fishing line.

Another trip to Streaky Bay in late 1953 saw him do it again with a white pointer weighing 1,076 kilograms (2,372 pounds) followed by another weighing 1,150 kilograms (2,536 pounds). There was no stopping him; he broke record after record, always his own, but the current world record shark wasn't hooked until April 1955.

Dean was fishing out of Ceduna, to the north-west of Port Lincoln, when he spotted the 'big one'. It was baited carefully, hooked and played for about seventy minutes, during which the shark went deep, but made a few scary surfaces when it could have snapped the line. But it didn't, and so it was pulled alongside the boat and then towed to Denial Bay, where the harbour master conducted the official weigh-in. The shark was 5.2 metres (17 feet) long and weighed 1,208 kilograms (2,664 pounds), a formidable beast. Since then, nobody aside from Frank Mundus has caught a white shark that has come to within 340 kilograms (750 pounds) of Alf Dean's record holder.

Even so, Dean was convinced there was an even bigger fish out there to catch. Back in 1952, he hooked a white pointer that he estimated to be close to 6 metres (20 feet) long and weigh in excess of 1,800 kilograms (4,000 pounds). He had the fish close to the boat, but his reel jammed, and the line snapped. He saw the same shark again several years later, but he couldn't get it to take the bait – once bitten, twice shy. Even so, at one stage, Alf Dean held seven world records, and all his sharks were caught between Kangaroo Island and the Great Australian Bight. It was a South Australia success story.

In the north-east, Hervey Bay – the whale-watching capital of Australia – is home to shark hunter Vic Hislop. Very different in character from Alf Dean, Hislop is more in the Mundus modus operandi. Unlike other shark hunters, who nowadays see sharks as an essential part of the ecosystem, Hislop described them as 'God's mistake' and that he felt compelled to right that mistake by killing them. As a result, he seems to have had a vendetta against sharks, white sharks in particular, and he's caught some pretty big ones. In 1987, for instance, he landed what was claimed to be a 6.3-metre (21-foot)-long white pointer and in 1985 a white shark of 6.6 metres (22 feet). Neither are reliable measurements. The sharks he caught were strung up outside his 'shark museum' until the neighbours couldn't stand the smell any more. The museum closed in 2007.

Hislop, however, was the shark hunter responsible for catching Damien Hirst's sharks. Hirst is a British artist who was a leading member of Young British Artists. Hislop supplied Hirst with twelve, one of which was sold for an undisclosed amount but estimated to be more than $8 million in 2004 as part of the artwork entitled *The Physical Impossibility of Death in the Mind of Someone Living*. It was commissioned by Charles Saatchi in 1991. The installation consists of a steel and glass tank filled with a 5 per cent formaldehyde solution in which a tiger shark, rather than a white shark, was preserved. To fit in with the theme, Hirst wanted something 'that could eat you'. Due to deterioration, the original shark was replaced by another which Hislop caught off the Queensland coast. Other works featuring Hislop's shark catches include Hirst's *The Immortal*, which is another steel and glass tank containing a white shark. There are several more.

Hislop himself tried to sell a preserved 5-metre (16-foot)-long white shark from his defunct museum for $30,000. It was listed on Gumtree, but a sale did not go ahead because the Australian authorities objected, so it was removed from the website. What happened to it thereafter is not known.

Of the internationally less well-known shark hunters, one that stands out is Russell J. Coles of North Carolina, USA. He was an avid fisherman and had his first encounter with a white shark in July 1902 in the Bight of Cape Lookout. He was using a stout trolling rod and a tarpon reel with 180 metres (590 feet) of line, and he hooked, played and landed a white shark after quite a struggle. It was dispatched with twenty shots from a heavy revolver, and was so big that six men could not lift it. Block and tackle had to be used, but even that failed at first.

The event reached the local newspaper, the Raleigh *News & Observer*, which reported that Coles had landed 'the largest shark that was ever caught in these waters'. It was measured at 5 metres (16 feet) long. It was enough for Coles to become obsessed with fishing for sharks, especially large ones. However, after he caught a shark so large with rod and reel, he said that he could not hope to equal it again. So he discarded that method and opted for other ways in which to catch and kill his quarry, some of them exceedingly bizarre.

The following year, for example, he was in a small skiff, harpooning sea turtles, when a large shark appeared. It was, he wrote later in a 1915 edition of *Proceedings of the Biological Society of Washington*, 'a huge shark of more than 20 feet [6 metres] in length', and it 'appeared alongside, within reach of my hand. It apparently had no fear of us, as it struck the side of the skiff with some force.'

Coles was well armed with a high-powered rifle, a light harpoon and a heavy knife. In a scientific paper in *Copeia*, dated 7 May 1919, he recalled the same event again, but with a few more details and a white shark that had shrunk somewhat.

> An 18-ft [5.5-metre] shark, easily recognizable as this species, charged, halting only when in contact with my skiff, where, with its large staring eye watching my every move, it lay for some seconds almost motionless with part of its back exposed above the surface, while I crouched with finger on the trigger of the high-powered rifle, aimed in front of the first dorsal fin. The shark then began a series of rapid evolutions, turning several times on its back while circling the skiff, into which it splashed much water.

Coles's shark then withdrew to about 100 metres (110 yards) away and just as it began a high-speed charge towards the boat, something quite unexpected occurred:

> Suddenly, in the line of its attack a large logger-head turtle came to the surface and was seized in the jaws of the shark, which I heard crushing through the shell of the turtle. I am convinced that this shark had satisfied himself that I was suitable for food and had only retired to acquire speed for leaping into the skiff and seizing me, and that the coming to the surface of the turtle at that instant was all that saved me from a dangerous, knife to sha-green fight. The shark seized the turtle in its jaws, and both disappeared below the surface. The next day I harpooned this turtle and found the upper shell for a width of nearly thirty inches [76 centimetres] showed the marks of the shark's teeth. The edge of the shell and the right hind flipper had been torn away.

Many of Coles's encounters with white sharks, which he referred to as 'man-eaters', were with very young individuals. In the *Copeia* paper, he describes harpooning two youngsters on separate occasions and carefully measuring them – one a male 1.8 metres (6 feet) long and the other a 2.2-metre (7.3-foot)-long female – and he enters into so much detail that he even describes the colour of the muscles.

> In colour the flesh was distinct rich light pink salmon (I have never seen the flesh of any other shark so coloured), except that extending along in the pink flesh on each side of the verte-bral column from skull to just above vent there was an almost round strip of nearly black flesh.

Coles then, and highly unusually for a scientific paper, goes on to describe the taste.

> Both pink and black flesh were eaten and proved excellent. Usually, the flesh of sharks is almost free of oil, but that of this fish was rich in oil, and its liver richest in oil of any that I have seen. It was the very finest shark, or, in fact, fish of any kind that I have ever eaten, its flavour being quite similar to a big, fat white shad. I made an entire meal of man-eater shark, eating nearly two pounds [900 grams] for dinner.

The following day he harpooned more sharks, including two more young female white sharks. At about the same time, he heard from other fishermen about a much larger white shark, 'as long as their 25 foot [7.6-metre] launch'. It was caught in a net and with much thrashing about, the fishermen eventually cut it free. Not long after, Coles came across the same shark trapped in nets for a second time. This time he was able to take a close look.

'My carefully noted observations justify the following claim of dimensions for it: length, 22 ft [6.7 metres]; head, larger than 50-gallon [190-litre] barrel; mouth, 3 ft. [90 centimetres] wide, weight, over 2 tons.'

He then went on to make an interesting observation. 'I consider it highly probable that this large shark was the mother of the young ones taken, and that she had given birth to them near Cape Lookout in May. These are points which make the presence of this species here still more interesting.'

Coles was not just any old shark hunter, but an inquisitive fellow who was making important scientific observations at the beginning of the twentieth century, when shark research was in its infancy. Many of his notes were sent on to the American Museum of Natural History.

A second large shark encounter, however, sounds extremely hairy to say the very least. This time, Coles was actually in the water with the shark, something he thought, for some inexplicable reason, was safer. He seemed to think that because he had become adept at using a large and lethal knife he was somehow protected.

> After trying for an hour to approach within harpooning distance of a large man-eater which was swimming in shallow water near the scene of my former encounter, I got over-board in a depth of five feet [1.5 metres] of water and had the boat retire to a distance of a hundred yards [90 metres] with the coil of rope, which was attached to the harpoon which I had with me. I also took with me half a bushel [18 litres] of crushed and broken fish to attract the shark, which was then swimming on or near the surface, half a mile [800 metres] to leeward of me.

The shark, like any self-respecting predator, zig-zagged upstream to follow the trail of scent towards Coles. He stood in the shallow water with harpoon poised.

> The shark charged from a hundred feet [30 metres] away with a speed which has to be seen to be appreciated. I met the onrushing shark by hurling my harpoon clear to the socket into it, near the angle of the jaw, and, as the iron entered its flesh, the shark leaped forward, catching me in the angle formed by its head and the harpoon handle, which caught me just under

the right arm, bruising me badly, while my face and neck were somewhat lacerated by coming in contact with the rough hide of the side of its head.

It was time to use the knife, but the force of the impact snapped the harpoon, and the shark made off. 'I never again employed the same black-smith to forge my harpoons,' he declared, 'but that poorly-made iron surely brought to a sudden ending a most exciting situation.'

Let's face it, he was lucky to be alive.

For more memories of Russell J. Coles, you can read about him on the website of historians David Cecelski and his daughter Vera, including Coles's other obsession with manta and mobula rays. *Shark Hunter: Russell Coles at Cape Lookout* is a fascinating series of articles on an extraordinary man.

Bringing our review of shark hunters up to date, we couldn't leave out Chip Michalove, a South Carolina charter boat captain who works out of Hilton Head Island to the north-east of Savannah. Michalove tells how he catches white sharks on almost every outing, and he uses careful observations of the area to be in the right place at the right time to intercept them. On his wall at home, for example, he has charts showing the sea surface temperatures when he catches white sharks or when not, what he thinks they are eating and what water clarity he believes they prefer. Armed with this data he has been working out the sharks' migration routes through South Carolina–Georgia waters.

And rather than string sharks up at the quay for people to gawp at, which is now illegal anyway, Michalove tags and releases them. He's so good at catching white sharks, scientists from the Atlantic White Shark Conservancy kit him out with a regular supply of satellite tags.

This fascination with sharks was instilled in Michalove from a very early age. When he was just four years old, his parents took him on a fishing charter with the legendary Fuzzy Davis out of Hilton Head. Davis caught a large shark and, although the young boy was terrified of it and he cried, it clearly made a lasting impression. At high school, he was still obsessed with white sharks, and he came up with a notion that nobody had thought about before. He noticed that large sea turtles were being washed up on South Carolina beaches with great semicircular bite marks and half-moon chunks taken out of them. People blamed tiger sharks, but the young Michalove thought the bites too big, and besides the carcasses appeared when tiger sharks had moved farther south for the winter. The only creature capable of making such sizeable bite marks, he thought, was the white shark.

After leaving school, Michalove's challenge to himself was to catch a white shark, but everybody thought he was in cloud-cuckoo-land. There are no white sharks around here, they said, but he proved them wrong, and when he did catch one, the first fisherman in South Carolina to do so, nobody believed him. The newspapers felt his story to be too far-fetched and wouldn't print it. Now, he catches and tags them almost every day.

One of his tagged white sharks, a 4-metre (13-foot)-long female with the nickname Leebeth, went on an extraordinary journey. She was released with her tag on 8 December 2023 and by July the following year she had travelled the width of the Gulf of Mexico, to the coast of Mexico, and back, northwards up the east coast of North America to Nova Scotia and entered the Gulf of St Lawrence, a total distance of 8,370 kilometres (5,201 miles) in less than eight months.

And the fisherman who sparked Michalove's love of sharks – Fuzzy Davis – has continued to stimulate young minds during the past few decades with a remarkable community programme for special needs children and adults called 'Fishing with Friends'. Every October for the past twenty-seven years or so, Davis has persuaded more than fifty charter captains on Hilton Head to donate their time and their vessels for a day of fishing. The best part, says Davis, is seeing the smile of children who have never been on a boat before catching their first fish.

CHAPTER 12

VULNERABLE SHARKS

White sharks are in danger, but nobody knows just how much danger because nobody knows how many remain. The feeling among shark scientists, however, is that populations of white sharks have been declining globally. The actual numbers generally don't exist, aside from a couple of exceptions. The data is just not available, but the Red List status of the species, according to the International Union for Conservation of Nature (IUCN), is 'vulnerable', one step down from 'endangered'. This means that the species is threatened with extinction unless the circumstances threatening its survival and ability to reproduce substantially improve. The main reason it has been assigned to this category is the fact that white sharks are caught accidentally by commercial and artisanal fisheries, and deliberately and illegally in some places. Global abundance appears to have dropped to 63 per cent of pre-1970s estimates, but this doesn't apply to white sharks everywhere. Scientists are keen to identify evolutionarily significant units because their management is crucial for white shark conservation as the species is facing various but often region-specific, man-made threats.

Off the coasts of Europe, which includes the Mediterranean region, for example, things are a darn sight worse. The species here is considered by the IUCN to be 'critically endangered', which means white sharks are facing the real possibility of extinction in the wild in the Mediterranean Sea and north-west European waters. Sharks here are caught accidentally on pelagic long lines, by bottom trawls and in drift nets and purse seine nets. There hasn't really been a white shark fishery in the Med, but sharks are killed after shark bite incidents and the ensuing media attention. White sharks also readily approach boats, take hooked fish and scavenge from nets with the risk of entrapment, particularly in the almadraba and mattanza nets for bluefin tuna, and a new hazard has arisen in recent years in the form of bluefin tuna farms, such as those off Malta, Spain and Libya. Research in Australia revealed that there is a minimal risk of white sharks being attracted to these net cages, despite the concentration of easy-to-catch fish. Nevertheless, a white shark chomped its way into a tuna enclosure off Tripoli, swam around for a couple of hours and then left, probably a little plumper than when it arrived. The crew, fortunately, did not kill the shark, but it could just as easily have gone the other way. Of nine confirmed tuna farm cases over five years in Australian waters, six sharks were killed, while the other three were already dead in the enclosure. However, more recently, fishermen are finding ways to release sharks, especially in the Mediterranean.

Even so, in the Mediterranean region, the list of potential hazards goes on. Degradation of the habitat has affected prey populations, such as monk seals and wild bluefin tuna, and human activities, such as mining and quarrying, oil and gas exploration, renewable energy systems and the concentration of shipping lanes all overlap with white shark living spaces.

The Mediterranean is a unique semi-enclosed sea, the

largest and deepest in the world. Its continental shelves are relatively narrow, so most of it is actually relatively deep. Yet people, down the centuries, have had a bigger impact on ecosystems here than in any other sea in the world. Young white sharks, for example, are more likely than older individuals to be the bycatch of long-line, drift-net fisheries or tuna traps, but 70 per cent of bycatch involving white sharks is due to purse seine fisheries. The purse seine net is a particularly difficult obstruction for white sharks to negotiate. It consists of a large circular net that gradually tightens to trap the fish inside. Once trapped, it's difficult for a white shark to escape.

The problem in the Med is that in some parts few people care, which is really a threat to the survival of white sharks, and other major apex predators, such as bluefin tuna, and to the survival of the fishermen who catch them. On most northern coasts fishing is regulated by the European Union and it's forbidden to catch white sharks or, if caught accidentally, they must be released immediately. On southern coasts, such as off Tunisia, things are very different. Sharks, including the occasional white shark, are caught without compunction; in fact, the World Wide Fund for Nature in north Africa suggests that 20 per cent of shark catches are deliberate and illegal. And we know it's true because the fishermen 'show off' their shark catches on social media. Fins are said to go to Asian restaurants in Europe, despite a ban, and shark meat is a traditional ingredient in local cuisine, although much is transferred to Sicilian fishing boats and sold in Italian markets where the meat commands higher prices. It's labelled 'swordfish'. The result, of course, is that populations of white sharks in the Med are hanging by a thread.

And it's not just in the Med. The pattern is repeated in many parts of the world's oceans, but not everywhere. In some places, such as the north-east coast of North America, the changes in attitude towards potentially dangerous sharks, a return to a historical feeding regime, in other words catching seals, and a ban on hunting white sharks in US waters since the 1990s has resulted in a slight but noticeable increase in numbers visiting the region. Actual numbers, again, are unknown. NOAA Fisheries simply states that 'the stock status for white shark populations in US waters is unknown and no stock assessments have been completed' but goes on to say that 'abundance trends have been increasing in the northwest Atlantic' and 'the northeastern Pacific white shark population appears to be increasing', which might explain why the white shark is not included, despite its IUCN status, in the list of endangered or threatened species under the USA federal Endangered Species Act (ESA).

Scientists moved to petition the authorities with data that proposed that the north-east Pacific population, at least, should be included. A biological review was organised by the NOAA Marine Fisheries Service, which included a range of fisheries biologists. The threats they recognised were not unlike those seen elsewhere: '1) fisheries mortality in US, Mexican and international waters, 2) loss of prey due to over-harvesting, 3) small population effects (such as human-caused wildlife disturbances), 4) disease and predation, 5) habitat degradation linked to contaminants, and 6) global climate change.' However, even though the north-east Pacific white sharks were recognised as a distinct population, it was deemed that the species need not be listed under the ESA at that time. That was in 2013.

Nevertheless, white sharks have been protected in federal waters since 1997, and state designations preceded or followed in previous and subsequent years. In fact, the first country to protect the white shark was South Africa in 1991. California was also ahead of the game and has protected its white sharks since 1994 while Massachusetts followed in 2005. In these waters, white sharks cannot be retained, even if they have been caught accidentally. If caught by sports fishermen, the sharks must be released immediately, no discussion.

In Mexico, the white shark has been protected since 2010 and in January 2023 it took the surprising step of banning shark tourism activities, including cage diving, film production and live-aboard diving, around Guadalupe Island, which had been until then one of the primary dive sites in the world to see exceptionally large white sharks and was host to an annual round of productions for Discovery Channel's Shark Week. The Mexican authorities said that the ban was in response to bad practice in the industry, such as mishandling the bait to attract white sharks, and tourists and film crews swimming outside cages and dumping pollutants at sea. Events they quote include the way, in 2016, a white shark became trapped among the bars of a cage and sustained serious injuries and a similar event in 2019 which probably resulted in an animal's death. The ban aims to protect not only these sharks, but also the whole Guadalupe biosphere reserve. It also effectively wipes out the entire cage diving industry, with the fear that without the constant presence of dive boats and their crews, poachers will move in and start to harvest the sharks illegally for their jaws, fins and meat. A set of jaws here, for instance, sells for about $5,000, according to one of the local operators. It would negate the whole reason for introducing the ban.

This is precisely what has happened at Malpelo Island, 500 kilometres (300 miles) west of the Pacific coast of Colombia. The sheer and barren rock is at the centre of the largest no-fishing zone in the eastern tropical Pacific and, in 2006, was declared by UNESCO as a natural World Heritage Site because of its status as an important shark reserve. Yet the sharks are not safe. Fishing boats from other Caribbean and South American countries and even from China come to the area to fish illegally. The Colombian navy catches some poachers, but only when out on narcotics patrols, so it has been left to a voluntary body – Biological Diversity Colombia – to patrol the seas around the island with their catamaran *Silky*. Without weapons, they try to persuade poachers to leave and it's working to some extent, but the Malpelo example does show how vulnerable the white sharks of Mexico's Guadalupe Island might be in the near future.

Internationally, white sharks have some degree of protection by being listed on Appendix II of CITES, a treaty that supposedly shields endangered plants and animals from the threats of international trade. It includes not only the whole fish, but also its body parts, such as jaws, teeth and fins. Another international agreement is the Memorandum of Understanding on the Conservation of Migratory Sharks, which recognises that white sharks enter the waters of several different countries, some geared up to protect the species, others not.

The white shark, after all, is a global species so it is exposed to the strictness or vagaries of each country's laws.

In Australian waters, for example, white pointers have been protected since 1999. The government recognised that populations were in decline due, in part, to sports fishing and sharks becoming trapped in nets protecting beaches. In 2002, a White Shark Recovery Plan was instigated, which called, among other things, for more research to determine where the sharks are at any one time and where they tend to go. In 2012, the existence of the two genetically distinct populations was realised – one in the east and the other in the west, the two separated by the Bass Strait – and so a revised plan was submitted in 2014. Even so, sharks are still killed by people here. There are accidental and illegal catches by commercial fisheries, and white sharks are still becoming trapped in beach protection nets. At one time, drumlines were deployed along with the nets to prevent a shark from reaching recreational bathers and surfers on beaches. These drumlines consisted of a baited hook, attached to a buoy and anchored to the seabed. Sharks would be caught and, if they didn't die on the hook, they were not relocated and released alive but shot, so shark numbers, including other species of large sharks, such as tiger sharks, have declined over the past half-century, according to the Commonwealth Scientific and Industrial Research Organisation (CSIRO), Australia's main scientific body.

Today, things have changed a little, and for the better. The drumlines still exist but they have been updated. Western Australia and New South Wales use SMART (Shark Management Alert in Real Time) drumlines, which are a step up from the old-fashioned ones. These consist of two buoys, a baited hook, a triggering magnet and a GPS communications unit, all of which is anchored to the seafloor, about 500 metres (550 yards) offshore. When a shark takes the bait, the magnet releases and a signal is sent to the drumline contractor who can then respond quickly to tag, relocate and release the shark. Queensland uses similar Catch Alert Drumlines (CAD). Both types of line are meant to be non-lethal.

Previously, back in 2014 after a spate of fatal white shark attacks, Western Australia had trialled conventional drumlines, which were far from non-lethal (see also Chapter 13). Many white sharks died unnecessarily. Then, the authorities trialled SMART drumlines in the Capes region with mixed results. The idea behind the project was to look at the movement patterns of white sharks after their release in order to determine whether relocation helped keep the beaches safer. It also assessed the impact of SMART drumlines on the mortality of white sharks and other hooked species.

During the 27-month trial period, 352 fish were caught, and 91 per cent of them were released alive and in good condition. Only two white sharks were caught, white sharks being the target species, and when they were released, they moved offshore and remained there for the rest of the experiment. However, twenty-four other white sharks were detected in the area, which just goes to show that Western Australia's white sharks are not amenable to capture. The conclusion was that one solution doesn't fit all, and that local conditions come into play. What might work in New South Wales does not necessarily work in Western Australia; and even in New South Wales, the new regime was not without its flaws.

Here, catches during the 2022–23 meshing season added to the overall population decline in white sharks when twenty-four target sharks (white sharks, tiger sharks and bull sharks) were caught, of which eighteen were white sharks, but only eight released alive. Most of these were immature, and their arrival in the meshed areas from October to December coincided with the most catches. Tagging studies have shown that the larger, adult sharks remain some way offshore and rarely become trapped in the nets or caught on drumlines, which is one blessing.

Another is that good practices were trialled and imposed, even during Western Australia's indifferent experience. The SMART drumlines here were positioned so that a vessel could reach the shark within thirty minutes of it being hooked, which gave it a good chance of surviving, being relocated and released. Humans benefited too. Beach users could access a SharkSmart Activity Map and the Surf Life Saving WA X feed for up-to-date news of captures, white sharks in particular.

A recent shark-spotting revolution is the use of drones, and they have been trialled at many beaches in Queensland, for example, with great success. Flights are limited by the weather, but, when conditions are good, large and potentially dangerous white sharks can be followed, and the beaches alerted to their presence. Approval rating from local residents is high. Up to 98 per cent of respondents to community surveys support the initiative, and 75 per cent of them are likely to select a beach where drones are flying.

Another significant development in Australia is a world first. Here, CSIRO researchers, working with other Australian academics and scientists from New Zealand, have used genetics to work out the number of white sharks in Australian waters. The technique is called close-kin mark-recapture and it involves taking a tissue sample from a living or dead shark, obtaining a genetic profile of the individual, and then comparing that profile to those of other sharks to determine if and how these sharks are related. Using this method, the research team has estimated that the total population is 8,000–10,000, of which about 2,210 are adults, the rest being newborns, young of the year, juveniles and subadults. As to whether the population is increasing or decreasing, it appears to be neither; the numbers are flat, which means Australia's white shark population might not be declining after all.

In New Zealand, white sharks have been protected within 370 kilometres (230 miles) of the coastline since 2007. The maximum penalty for taking a white shark is up to six months in prison along with a maximum fine of $250,000, and this also applies to New Zealand-flagged boats operating outside the zone. Researchers have estimated that between 1,000 and 5,000 mature individuals reside in New Zealand waters, the vague figures due to what has been called 'Data Poor'.

In South Africa, the data is not poor, but it is somewhat disturbing. On the Western Cape, two areas have been traditionally white shark hotspots – False Bay and Gansbaai – the so-called 'white shark capital of the world', with a thriving cage diving industry… but not any more. In just eight years, white sharks have almost totally disappeared. Studying them has been an international team, including researchers from Oregon, Rhodes and Stellenbosch universities, which has been tagging South Africa's white sharks for close to twenty years, but a recent development has had them perplexed. As many as eighteen of twenty-one white sharks tagged in Mossel Bay since 2019 have gone. They don't appear to have moved elsewhere, to the east, for example; they've simply vanished. It was thought that they might have fled from the two infamous orcas, Port and Starboard, that began terrorising shark communities in 2015 (see also Chapter 8), but, according to a University of Exeter scientist, the start of the decline predates them, starting in Gansbaai in 2013.

Before that, back in 2010, there were thought to be between 500 and 1,000 white sharks off South Africa. By 2013, a study revealed close to 900 and in 2016 research at Stellenbosch University estimated between 353 and 522, but in recent years, not only have the tagged sharks gone AWOL, researchers have also failed to spot the presence of large, mature females – those longer than 4 metres (13 feet) – even when there are fresh whale carcasses to attract them in. They're just not present, a sure sign that the population is not doing well. If there are no mature females, then there will be no new pups, another significant contribution to population decline.

This is not helped by the way the KwaZulu-Natal Sharks Board catches and kills white sharks with beach protection nets and drumlines, in the mistaken belief that culling sharks safeguards bathers and surfers on South African beaches. Even though white sharks have been protected in South Africa since 1991, the mortality from beach defences is staggering. During the four decades from 1978 to 2018 drumlines and shark nets caught 1,317 white sharks of which 1,108 were killed; that's roughly twenty-eight white sharks killed every year for the past forty years. Then, there are the fisheries.

Despite catching white sharks being illegal, a legal long-line fishery that targets smaller shark species living on or close to the seabed inevitably has the larger sharks sniffing around. A study of stomach contents of South African white sharks revealed that they often consume bronze whalers, smooth-hounds and soupfin sharks, precisely the target species of the long-line fishery. Many are caught incidentally, a conservative estimate being forty white sharks a year between 2008 and 2019, and those are only the ones reported. Photographer Oliver Godfrey went out on one of the fishing boats and witnessed how white sharks were hooked, killed and thrown back in the sea, and nothing was officially reported, which means the forty sharks a year really is a conservative estimate. This, though, is a double whammy because not only are the white sharks themselves being killed, but their primary food species are also being removed.

Chris Fallows of Apex Shark Expeditions witnessed the decline firsthand. He tells of how the South African Department of Forestry, Fisheries and the Environment handed out permits to fish for small sharks in False Bay in the 1990s. False Bay was where white sharks go for eight months when they've run out of baby seals at Seal Island, targeting the same fish as the fishery. A small but noticeable decline in white sharks occurred at Seal Island, where Fallows embarks on cage diving experiences. Then, the fishing activity increased greatly, and the white sharks disappeared altogether. The multi-million-rand ecotourism industry collapsed, and the fishery is heading the same way. As the researchers themselves have said, 'South Africa could go down in history not only as the country to first protect white sharks, but also the first country to oversee their downfall.'

One option for a population of white sharks to recover naturally is for sharks from other locations to migrate in and fill the gaps. Alas, if it were only that easy. Scientists from Nord University in Norway have determined that there are just three genetically distinct lineages in the world. This split occurred 100,000 to 200,000 years ago and has been maintained to this day. The three groups are located in the north Atlantic, the Indo-Pacific region and the north Pacific respectively. However, they do not interbreed, which means that if one is approaching extinction or becomes extinct it won't be replaced.

Like salmon, female white sharks tend to return to the site of their own birth to drop their pups. It means that mitochondrial DNA, which is inherited from the shark's mother, is like a passport. It shows exactly from where the shark has come. Earlier studies have looked at shark DNA to determine genetic diversity, but a portion of that maternal DNA is subject to mutations. In a new study, researchers have looked at variations in that DNA at a single-nucleotide level, the basic building block of DNA. From eighty-nine white shark samples from around the world, the scientists were able to group related genetic sequences using a statistical algorithm and were left with three distinct populations.

It's thought the reason for the split – during the Penultimate Glaciation Period – was that sea levels were considerably lower than they are today, by as much as 80 metres (260 feet), which probably resulted in shifts in ocean currents and sea surface temperatures, creating biogeographical barriers between the groups. Genes are not seen to move across boundaries and so there must be some kind of natural selection taking place by which the three lineages have become adapted to the places that they're in.

There was, however, one instance of interbreeding. A shark living in what has become known as the 'Bermuda Triangle' in the north-west Atlantic was a hybrid of the Indo-Pacific and north-west Atlantic groups. Why is a bit of a mystery, although hybridisation may have occurred more frequently in the past and the lines died out due to natural selection because the offspring were not well enough equipped to survive the prevailing conditions. It indicates that if a population should be reduced and another strays into its territory, then interbreeding might well result in offspring that don't survive. It's not good.

So, how about captive breeding as a means to boost numbers? It seems to be working with other large species, such as bluefin tuna that are raised on farms, so why not white sharks? The short answer is that it's a non-starter. White sharks do not fare well in captivity. In 2016, for example, a 3.5-metre (11.5-ft)-long male white shark was caught alive in a net by fishermen off the south-west coast of Japan, and it was donated to the Okinawa Churaumi Aquarium. It was placed in the 'Sea of Dangerous Sharks' exhibit, where it swam about and appeared to be doing well, but after a while, it was seen to swim erratically, drop to the floor of the tank and not feed. Within three days, it was dead.

Previously, Monterey Aquarium had some success in keeping white sharks. Their sharks were much smaller than the Japanese specimen. They were also acclimatised in open pens before being introduced to the exhibition tank, and, when they began to look stressed or unwell, they were returned to the sea. Short-term spells in captivity like these became the industry norm, with Monterey Aquarium keeping white sharks in a huge million-gallon (3,800-cubic-metre) tank for 198, 137 and 152 days in 2004, 2006 and 2007 respectively. All were successfully released. However, the first attempt was at the Steinhart Aquarium in San Francisco.

The date was 19 August 1980, when Tomales Bay fisherman Al Wilson caught a white shark in the neighbouring Bodega Bay and kept it alive while hauling it back slowly to port. It was the second specimen that Wilson had caught, the previous one having nearly killed him. It was about 6 a.m. when he first called SeaWorld in San Diego, which was offering $7,000 for a white shark, but nobody answered, so he called the Steinhart, which was offering $500 for the first two days, and $100 for each day the shark was exhibited, to a maximum of $5,000. A young researcher answered, which triggered plans the aquarium had made for just such an eventuality. They had tried with an ailing white shark previously, and it had died within a day, but this time they were ready with a special truck, a tank and a canvas sling with which to lift the shark, and the quick-thinking curator, John McCosker, thought the 2.1-metre (7-foot)-long juvenile female might survive in the aquarium's large circular tank. This would allow the shark to swim continuously and ensure water passed into her mouth and over her gills; and, indeed, the shark, which Wilson had nicknamed Sandy, made herself at home. She was the first to survive in captivity, and it certainly caused a commotion. The *San Francisco Chronicle* had it as headline news for several days and people queued around the block to come and gawp, some waiting in line for more than three hours. Sandy was so popular that there were record admissions to the aquarium that week.

However, this extraordinary phenomenon didn't last long. Sandy became stressed by the bright lights, and every time she passed a certain point in the tank she would either try to turn around or sink to the bottom, and a diver had to go in with her to keep her moving. Electronics engineers discovered that at that point there was an electrical anomaly which the researchers could barely detect, but the shark certainly did. By day four, though, McCosker decided they should return the shark to the ocean. She was tagged and transported to the Farallon Islands, where she was released. Tagging data shows she lived for at least another year before her tag ceased to transmit, but the experiments had limited success. Keeping sharks in captivity is not an answer to the white shark's population problem.

You will often hear shark scientists say that white sharks are significantly smaller than they used to be. In the past, there is no doubt in many minds that white sharks were bigger than they are today. It's not that they have mysteriously shrunk, rather it's the result of sharks not living long enough for them to grow to greater size, the result of being caught either deliberately or as incidental catch in commercial fisheries the world over. If white sharks are generally smaller than in the past, then fewer could be reaching maturity and having pups each year, another contribution to population decline.

Also, new techniques for determining the age of sharks have revealed that great white sharks do not live to a modest twenty-five or thirty years as was first thought but to the grand old age of seventy, and so the turnover of white sharks is small.

An adult female has between two and fourteen pups every two years, if she's lucky, and, as we learned in Chapter 9, she will not reach maturity until she is about 4.5 metres (15 feet) long or about thirty-three years old; males will do so at about 3.6 metres (12 feet) or twenty-six years. So, we have a slow-growing, long-lived species, having pups only every couple of years, so it is extremely vulnerable to hostile forces, such as overfishing, bycatch, coastal pollution and, yes, climate change, with shifts in the distribution of prey and changes in the migratory patterns of the sharks themselves, something which, as we have read, is already occurring on the west coast of the USA with juvenile white sharks moving increasingly towards the north.

If things do not improve substantially, all this new research indicates that, despite the upswing in the north-east Pacific and north-west Atlantic, the white shark's status could ultimately change for the worse: it could go extinct more quickly even than the tiger. Before he died in 2006, Peter Benchley himself remarked that 'the shark in an updated *Jaws* could not be the villain; it would have to be written as the victim; for worldwide, sharks are much more the oppressed than the oppressors.' How right he was.

CHAPTER 13

THE JAWS EFFECT

There is a creature alive today which has survived millions of years of evolution… without change, without passion, without logic. It lives to kill – a mindless eating machine. It will attack and devour anything. Try to imagine meeting the Devil… with jaws. The nation's #1 best-selling novel is now the year's most terrifying movie.

This was the tagline in the first trailers for the motion picture *Jaws*. Even before the film was released, it was already sensational. So, what has all the fuss been about?

Jaws is the story of a fictional white shark that terrorises a seaside town and its people. Events occur around the 4 July holiday season, when an influx of visitors descends on the town, ensuring that local businesses survive for another year. It's the worst possible time for a monstrous shark to turn up, and the local administration is torn between keeping people safe and filling the tills.

Let's take a look at the plot.

Steven Spielberg's motion picture, based on the book by Peter Benchley, is set in the fictional town of Amity Island on the southern shores of Long Island, in real life Martha's Vineyard, an island to the south of Cape Cod. It features in another of the film's taglines: 'Amity Island had everything. Clear skies. Gentle surf. Warm water. People flocked there every summer. It was the perfect feeding ground.'

The film itself starts with a teenager, played by Susan Backlinie, skinny-dipping only to become the shark's first and most memorable meal. Her remains are washed ashore the next morning and discovered by a deeply shocked policeman, Deputy Leonard Hendricks, played by Jeffrey Kramer. The medical examiner determines that she was taken by a shark. It prompts the police chief, Martin Brody, played by Roy Scheider, and newly arrived from New York City, to close the beaches. Amity's mayor, Larry Vaughn, played by Murray Hamilton, is horrified, not due to the presence of the shark, but because of the impact that closing the beaches would have on the town's summer economy. He persuades Brody to open the beaches, and the police chief reluctantly agrees. Under pressure, the local coroner concurs with the mayor that the girl was killed in a boating accident, and the truth is buried; that is, until a young boy is killed on a crowded beach, causing a mass panic and exodus from the sea. A bounty is placed on the shark's head and a fleet of vigilantes commandeers anything that floats in order to head out to the ocean to catch it.

At a town meeting, the eccentric veteran shark fisherman Sam Quint, played by Robert Shaw, indicates he will catch the shark for $10,000, at which point Matt Hooper, played by Richard Dreyfuss, a marine biologist invited to Amity by Brody, examines the first victim and confirms she was bitten by a large shark. One of the fishermen, meanwhile, has caught a large shark – a tiger shark, and has strung it up on the harbourside (it actually came from Florida and was pretty ripe during the filming). Mayor Vaughn declares that this is the shark responsible for the deaths and that the beaches are safe to remain open. Hooper, though, dissects the shark and finds no human remains in the stomach, just some semi-digested fish and a car number plate, so he points out that this was not the killer. The offending shark must still be out there, he says.

Hooper and Brody, who is actually afraid of the ocean, take Hooper's boat on a night excursion to search for the shark, when they come across the half-sunken boat of a local fisherman, Ben Gardner, played by real-life local fisherman Craig Kingsbury. Hooper puts on his scuba gear and dives down to see what's what. He swims to a great gash in the hull where he finds a large shark's tooth embedded in the wood. Unfortunately, he is startled by the corpse of the fisherman, probably the point in the film that scares most cinemagoers, and drops the tooth, but he held it for long enough to realise that their adversary is an exceptionally large great white shark.

Meanwhile, the shark, which has yet to be seen properly in the film, enters a lagoon behind the beach – in reality Sengekontacket Pond, separated from the Atlantic Ocean by the American Legion Memorial Bridge, known popularly as 'Jaws Bridge' – where the shark kills a rower and almost catches Chief Brody's son. Vaughn is persuaded to hire Quint and so he, Brody and Hooper go in search of the shark. While Brody is laying down a chum line, the shark makes its first big appearance with mouth agape. Brody turns to Quint and delivers the immortal line 'You're gonna need a bigger boat' – not in the script but an ad lib from Scheider. Spielberg's shark was 7.6 metres (25 feet) long and weighed about 3 tonnes/tons, a true giant among sharks, but, in reality, not that much bigger than known outsize white sharks (see also Chapter 10).

To cut a long story short, Quint is devoured by the shark and his boat so badly damaged that it begins to sink in shallow water. Perched in the crow's nest, Brody kills the shark by firing at a scuba tank in the shark's mouth and it explodes. Hooper, who had tried to kill the shark with a hypodermic spear laced with strychnine from a shark cage, surfaces alongside Brody and the two paddle towards the shore on a raft made from yellow barrels. Cue closing credits.

Surprising though it may seem, the film almost wasn't made. The producers, Richard D. Zanuck and David Brown, read the book overnight and agreed it would make a great film. Brown later claimed that if they had read the book a second time, they would probably have given up – it was too difficult. Nevertheless, it went ahead and after a period of fits and starts, when various directors were considered and the script was written and rewritten, Spielberg was the chosen one. This is when things began to unravel as the director wanted to film everything for real in, on and beside the ocean, rather than on a Hollywood back lot. Shooting this way is fraught with problems. The environment is forever changing – the light, sea conditions and, of course, the complete uncontrollability of virtually everything. The studio offered to build a large tank, so Spielberg would have had more control, but he was having none of it. He wanted it to look and feel real.

When the 200-strong cast and crew descended on Martha's Vineyard, true to form, everything that could go wrong went wrong. First up were the local regattas. Amateur crews were racing their yachts back and forth, so the film crew had to wait at each shot until they cleared frame. It wouldn't have helped the sense of danger in a scary movie to have these peaceful sailing boats in the background. Then, there were the currents that caused all the filming craft to drag their anchors and so be out of position. By the time they were manoeuvred back in place it was lunchtime and not a shot had been made. It was unbelievably tedious and frustrating, and the costs were mounting. It all now came down to the robotic shark working well.

Spielberg's monster shark was nicknamed Bruce, after his lawyer Bruce Ramer, but the three Bruces that the prop team constructed were technically difficult to operate in the sea; in fact, on the first day of camera tests, with Zanuck and Brown in attendance, the shark was supposed to breach, which it did, but then there was an explosion of bubbles and Bruce went to the seabed. To continue with the film, the director had to find more ingenious and less time-consuming ways to create the scary atmosphere that was to pervade the story. The result was John Williams's famous musical score, with the two chords E and F, and not seeing the shark but hinting at its presence for much of the film – it doesn't fully appear until one hour and twenty-one minutes into a two-hour movie. Thus, one of the most famous movies in cinema history hit the jackpot simply because of a mistake and the talents of a creative director who corrected it. 'What's scary about the movie', Spielberg told the *Times Saturday Review*, 'is the unseen, not what we see.' It was what John Ó Maoilearca, professor of film at the University of Kingston, described to the BBC. 'The enemy you cannot see – something intent on devouring you in an element you're unfamiliar with – is a long-held belief,' he said. 'When we look into the water we see our own faces staring back at us. We know there is life down there, but it's thriving in an environment that is alien to us. And possibly plotting our destruction.'

Our vulnerability in the ocean, suggests nature writer Richard Kerridge of Bath Spa University, particularly the fear of what might come up from underneath, enhanced the fear factor in *Jaws*. In fact, the psychological impact of *Jaws* on cinemagoers was nothing short of extraordinary. The film and its enormous rogue shark were so scary that psychologist Gabriella Hancock, working with biologists at the Shark Lab at California State University, Long Beach, reckons an entire generation suffered galeophobia – an irrational fear of sharks. This fear, she feels, is not innate. Many people's fear of sharks is probably learned and greatly influenced by stories in the popular media. Five-month-old babies, she says, are not the slightest bit afraid of sharks.

In fact, we need predators like sharks, says veteran entomologist and sociobiologist E. O. Wilson of Harvard University. 'We are not afraid of predators,' he's quoted as saying, 'we're transfixed by them, prone to weave stories and fables, and chatter endlessly about them, because fascination creates preparedness, and preparedness, survival. In a deeply tribal way, we love our monsters.'

Peter Benchley, the author of the book on which the film was based, felt the *Jaws* Effect was something even deeper. He acknowledged that it had 'inadvertently tapped a profound, subconscious, atavistic fear in the public, fear not only of sharks, but of the sea itself.' Indeed, on the release of the movie, people were suddenly afraid to enter the sea.

A former teacher from Wythall, Worcestershire, England, was an extreme example of this. She recalls how she saw the film when twelve years old (despite the scariness it was rated A, now replaced by PG, in the UK film classification system) and was terrified of going into the water for thirty years, making beach holidays a complete nightmare. Aged forty-four, she overcame her fear in an unexpected way, by learning to scuba dive and then swim with sharks, including a large bull shark, a hammerhead and a pack of 200 grey reef sharks. She was hooked, and wrote a children's book, *Sharks Are Scary Aren't They?*, based on her experiences. The idea was to help prevent children from having the same fear that she had. She hadn't been alone.

One contributor to a social media website tells how *Jaws* 'was the first movie that scared the daylights out of me. I hadn't seen anything so gory before. I remember my legs being shaky on the walk home, and I didn't go into the water at the beach that whole summer.' Another recalls how 'that movie completely ruined a childhood vacation to Florida for me. I was convinced if I couldn't see my feet, my legs would become shark snacks,' which was a common fear: 'If I can't see through the surface of the water,' one wrote, 'then there are definitely sharks waiting for me,' and another penned: 'I know it's so illogical, but that movie not only made me fear the ocean but swimming pools too.' An inhabitant of Newport Beach, California, saw *Jaws* on its opening day, and he didn't get into the water for a year. 'After seeing *Jaws*,' another recalled, 'I stayed out of the ocean for an entire summer out of sheer terror. I lived in walking distance of the beach, too, so I can easily say that this movie changed my life.' In the weeks after *Jaws* was released, beach attendance across the USA was demonstrably down.

It also had a considerable impact on the general public's perception of sharks in general and the white shark in particular, most of it negative and something that has persisted in popular culture to this day. The '*Jaws* Effect', as it has become known, boosted the notion that sharks intentionally attack humans, and this is further amplified in the popular media when reporting real shark bite incidents, influencing people's attitudes and perceptions such that it exaggerates the danger that these animals pose. The result is that white sharks are vilified.

It has also influenced greatly the way politicians view shark bite incidents and their response, when it is politically expedient to be seen to do something – anything! In the case of *Jaws*, it was to kill the shark, and in Western Australia between 2000 and 2014 a real-life series of fatal white shark attacks seemed to echo events in the film. This was the *Jaws* Effect.

Described simply, the *Jaws* Effect is the way people with some kind of political power and/or authority in society, who can have significant influence on decisions, policy, media coverage and outcomes, use fictional portrayals in film, such as in *Jaws*, as the basis to explain and deal with real-life events. It is, says University of Sydney researcher Chris Pepin-Neff in a scientific paper published in the *Australian Journal of Political Science* in 2014, 'a political instrument in policymaking that reinforces three themes: that sharks are intentionally hunting people, shark bites are fatal events and killing individual sharks will solve the problem'. The title of his paper was 'The *Jaws* Effect: how movie narratives are used to influence policy responses to shark bites in Western Australia', and it focused on the series of fatal white shark bite incidents between 2000 and 2014 and the furore that accompanied them.

To put this spell of shark bite incidents into perspective, between 1876 and 2002, there were just two confirmed fatalities and six unprovoked attacks on people by white sharks in Western Australia, which means that there was one death every sixty years and one unprovoked non-fatal attack in twenty years. You would be more likely to win the national lottery, it was said, than be mauled here by a white shark. However, in recent years, a heck of a lot more people are in the water – paddling, bathing, surfing, paddleboarding and kayaking – than there used to be, and as a consequence, Western Australia white sharks have been in the headlines multiple times, and for all the wrong reasons.

During the period in question, for instance, Western Australia experienced more fatal white shark bite incidents in a shorter time frame than has ever been reported in Australian history. Between 2000 and 2014, there were thirteen fatalities, with eight occurring since 2010, which, as you can imagine, has made it a big issue locally. To make matters worse, fatal bite incidents involving white sharks, during the same period, were reported elsewhere in Australia: seven in South Australia, four in Queensland and three in New South Wales.

This was Pepin-Neff's moment to take a look at the political fallout. His belief is that one of the first things from *Jaws* that policymakers seize on is the intention of the shark. The shark in the film deliberately hunted people, a throwback to Victor Coppleson's questionable concept of the 'rogue shark' (see also Chapter 3) which was doing all the killing. A second point is that the fatal shark attacks occurred over a short time in a relatively small area, as well as reinforcing the notion that white shark bites are inevitably fatal events. In the film, for example, a girl disappears while having a nighttime dip, a young boy playing on an inflatable lilo is killed, causing a mass exodus of bathers from the water, a fisherman is found dead in his sunken boat, a rower becomes a victim, and finally the character Quint is taken by the shark. In each case, the brutal situation becomes an essential element of drama in the film. Thirdly, there was the belief that the only way to resolve the crisis was for the two surviving co-stars to kill the offending shark. It was these three main fictional themes that translated into the real-life events in Western Australia.

At the start of this disastrous period, during the austral spring and summer of 2000, three fatal white shark bite incidents occurred within just three months – two off South Australia in September that occurred one day after the other, but geographically far apart. South Australians were up in arms and called for the shark or sharks to be killed, but South Australia law forbids this. Local fishing personalities added fuel to the fire by calling not only for a cull, but also to reverse the law protecting white sharks, because, they said, the shark will kill again – another reference to the rogue shark.

Two months later, on 6 November, the popular Cottesloe Beach near Perth, Western Australia, was the site of attacks on two swimmers taking a morning dip in waist-deep water. One survived, the other didn't. He died from loss of blood. One of the rescuers described the scene as like something out of *Jaws*. Another said she always thought the film *Jaws* was exaggerated, but not after what she had seen that day. Townspeople were openly comparing their situation to that of the fictional Amity.

Local dignitaries pontificated widely, stating the rogue shark must be killed, but after a week of unsuccessful hunts, the local populace was not unanimous in its condemnation. In a radio show, for instance, two-thirds of callers were against killing the white pointer. The voice of reason also pitched in. CSIRO shark scientist John Stevens said that humans were not normal prey items for sharks and, if they were, there would be considerably more shark bite incidents than we see today. But it cut no ice with the policymakers. Their view that the shark must be killed to safeguard people had become well entrenched.

At this point Peter Benchley entered the fray. He wrote an open letter to *The Guardian* newspaper. He began by saying how he couldn't pretend to comprehend the grief of the victims' family and friends and would not conceive of diminishing the horror of the attack, but then followed with:

> I plead with the people of Australia – who live with, understand and, in general, respect sharks more than any other nation on Earth – to refrain from slaughtering this magnificent ocean predator in the hope of achieving some catharsis, some fleeting satisfaction from wreaking vengeance on one of Nature's most exquisite creations.

He went on to point out that 'this was not a rogue shark, tantalised by the taste of human flesh and bound now to kill and kill again. Such creatures do not exist,' he wrote, 'despite what you might have derived from *Jaws*.'

Nevertheless, a special order aimed at killing sharks was

adopted as formal policy and contained in the Western Australia Shark Response Plan. It proclaimed that 'in the event of a shark attacking, or attempting to attack, a person, fisheries officers would, upon verification of the identity of the animal, immediately attempt to kill the shark.'

With no fatalities in Western Australia during 2001 and 2002, things went quiet for a while but, come 2003, *Jaws* raised its ugly head again. This time it was another shark encounter at the popular Cottesloe Beach, but one which didn't result in any injuries or deaths. However, it was enough to prompt more rhetoric in the Western Australia Parliament. According to Hansard, the official record of debates, reference was made back to the 2000 attack at Cottesloe Beach. In 2000, the shark should have been killed, it was recorded, and that the recent encounter was with the same shark. The rogue shark was resurrected yet again.

The year 2003 had no white shark fatalities, but on 10 July 2004, a 39-year-old man was savaged by what were thought to be two sharks while surfing at Lefthanders Beach, near Gracetown to the south of Perth. A sixteen-year-old surfing nearby told ABC News how he saw 'this big shark eating this guy, with another shark circling. There was blood everywhere and I saw the shark get him once more, then he was lying face down in the water, which is when we saw the other shark coming towards us, so I caught a wave in as fast as I could.'

On 19 March 2005, the Abrolhos Islands off the coast of Western Australia became the focus of attention when a white shark, estimated to be about 6 metres (20 feet) long, sliced a snorkeller in half. He died instantly. However, there were no plans to seek revenge. There was no shark hunt.

On 27 February 2008, a 51-year-old man was killed while snorkelling for crabs off a beach at Port Kennedy, 32 kilometres (20 miles) south of Perth, in a pretty harrowing attack. Witnesses described how the shark rolled over with the victim in its jaws. People on the beach looked on in horror.

Over a fourteen-month period during 2010–11, several more fatal white shark attacks occurred, including a 31-year-old man who died while surfing near Gracetown on 17 August 2010. There were seals in the area at the time. It was followed by a boogie boarder at Bunker Bay on 4 September 2011, another fatality at Cottesloe Beach on 10 October and a US tourist swimming near the popular Rottnest Island on 22 October. The response from the state government was pure *Jaws* Effect.

Immediately following the third incident the first kill order was issued. It was assumed that a rogue shark was responsible for all three incidents, so a Department of Fisheries vessel was dispatched to kill the shark. It failed. More fatalities followed.

On 31 March 2012, a diver from Perth was attacked by a white shark estimated to be about 4 metres (13 feet) long. He was 1.6 kilometres (1 mile) offshore of Stratham Beach near Port Geographe Marina, 230 kilometres (143 miles) south of Perth. Inevitably, this triggered the *Jaws* Effect again. Local beaches were closed. The elite of Busselton (the nearest town) called for a cull, suggesting that the attack 'would damage tourism in a popular area'. The City of Busselton president told *The Advertiser*:

I think there should be a culling programme because it's gone too crazy. How many more of these tragic deaths can we continue to have? It's far too many. I know people disagree with me, but I don't care; this is my opinion. If they have attacked humans, they should be destroyed.

It was assumed by some of the more informed people in authority that the guilty shark would have left the area, so they did not issue a kill order. Indeed, the Western Australia premier issued a statement:

I think everyone who goes in to the water knows this is the habitat of the shark and there is a certain amount of risk, particularly if you are a long way off the beach or particularly if you are around reefs and diving and the like. But if it was, for example, closer to the shore where people are, then consideration would be given to a kill order.

Unfortunately, the deaths continued. On 14 July 2012, Wedge Island, some 135 kilometres (84 miles) north of Perth, was the location for a surfer's demise. He and a friend were surfing at a bay known locally as Dolphins. They had been surfing for an hour and were resting on their boards when one of them noticed a white shark below, close to the seabed, a typical white shark attack position. The two scrambled for the shore, but the shark surfaced and headed for one of the surfers who was kneeling on his board. He tried to kick the shark, but it grabbed him. Two men on a jet ski saw the attack. One took the other surfer into shore and went straight out again.

'There was blood everywhere' he said, 'and a massive, massive great white circling the body.'

The jet skier tried to get close to the inert body, but the shark attacked his machine. 'I reached to grab the body, and the shark came at me on the jet-ski and tried to knock me off, and I did another loop and when I came back to the body, the shark took it,' he said.

On 23 November 2013, a surfer was killed near Gracetown, the third fatality in the area in recent times. The shark had made contact with another board before attacking that of the deceased. He was eventually pulled from the water and eye-witnesses said one of his arms was missing and flesh off one of his legs. Officers from the Department of Fisheries issued an 'imminent threat danger' notice and turned up to catch the shark. Needless to say, they didn't succeed. Celebrity chef Heston Blumenthal had swum in the same waters a few hours beforehand.

By this time, every attack had become immediate international news, so the Western Australia government was under enormous pressure to do something. Its response in 2014 was to order a cull using drumlines. Any shark caught that was more than 3.5 metres (11.5 feet) long – and this would include white sharks, tiger sharks and bull sharks – was shot. Newspaper reports describe, for instance, how a large female tiger shark was shot four times with a .22 calibre rifle and her body dumped into the ocean. The government, however, underestimated the feelings of local communities, the controversial move prompting thousands of people to take to the streets in protest. The crisis was brought to a head when Australia's Environment Protection Agency stepped in and questioned the scientific validity of the cull and promptly blocked it. The problem the west coast communities were left with was the negative impact on their tourist industry.

By the end of summer 2014, when there were normally

beaches packed with bathers and surfers, there were very few people about. On one of the state's prettiest beaches, the Cottesloe in the city of Perth, site of several white shark bite incidents, almost nobody was in the water.

'It's hard to convey to outsiders the impact in our community of these shark attacks,' wrote a local journalist in *The Australian* newspaper. 'People on the beach talk about sharks, people in the shops talk about sharks, patients and staff at the doctor's surgery talk about sharks. But few are sure just what if anything, should be done.' It was the worst nightmare of Amity's Mayor Lawrence 'Larry' Vaughn in *Jaws*, and something he was desperate to avoid.

In his learned paper, Chris Pepin-Neff summarised these incidents.

> These examples in Western Australia highlight how a fiction-based policy may serve political goals. Telling the story of a serial killing shark helped two different governments … maintain control of the narrative and achieve their policy outputs. It may also have provided an easier path because a more scientific-based narrative meant telling the public that nothing can be done or that the government did not know what was going on … this research identifies a worrying style of policy-making where widely known fiction can be used to navigate the attribution of blame and to prescribe policy responses.

But, alas, that was not the end of the fiction and of the slaughter. After the federal ban on extending the permanent drumline trial, politicians were still delivering the same old *Jaws* Effect line. The premier stated: 'I think our focus will be now what do you do with perhaps a rogue shark that stays in the area and is an imminent threat to beachgoers and I think that shark has to be destroyed and moved, I don't think it's acceptable.'

And true to his word, after a serious shark bite incident near Esperance in October 2014, a killing order was announced. A 23-year-old surfer was mauled by a white pointer, estimated to be up to 4.5 metres (15 feet) long, at remote Kelpids Beach in Wylie Bay. The victim lost parts of both arms, but he lived. Two drumlines were put out and a white shark was caught within an hour. Whether it was the perpetrator is anyone's guess, but that day two of these 'vulnerable' creatures were killed. Both had empty stomachs aside from the bait. There could be a couple of reasons for this: either they hadn't eaten recently, or they had regurgitated their stomach contents while on the hook, so it's hard to point the finger. However, it also means that the two might well have been killed unnecessarily, and all due to the *Jaws* Effect.

Since 2014, there have been at least seven fatal white shark bite incidents in Western Australia, including two more at Wylie Bay, in one of which a white shark was killed. Again, it's not known whether it was the shark responsible for the attack.

So, if all this is biological baloney and can be put down to the *Jaws* Effect, do we know what the real facts are about what has been going on in Western Australia? What information is available for people to assess their personal risk of being involved in a shark bite incident in Western Australia waters? Unfortunately, very little, but a few things are becoming clearer.

Peter Sprivulis, professor of emergency medicine at the University of Western Australia, believes there is a link between the west coast humpback whale migration and the prevalence of white sharks. He reported that shark bite frequency grew exponentially between 1973 and 2013. During the same time period, humpback whale populations have grown at roughly 10 per cent per year, so that's significant.

The whales head northwards along the west coast of Western Australia in early winter, reaching the Kimberley region where mothers give birth. In late winter and early spring, they head back south to the Southern Ocean. Mothers and newborn calves tend to lag behind the main migration, stopping off to rest in bays and inlets along the coast. So, it's no coincidence then that about two-thirds of white shark bite incidents occur when the whales are passing through, and roughly half are in spring when the mothers and calves are moving southwards. It has led an enlightened Department of Fisheries to state that, when assessing personal risk, several factors should be considered – depth of water, water temperature, distance from the shore, water activity, such as bathing, surfing or paddleboarding, and the presence of marine mammals, for instance whales and seals.

An analysis of the statistics by Sprivulis reveals that for bathers within 25 metres (82 feet) of the shore, during summer when whales are absent, the risk of being bitten by a white shark is one in 30 million for each swimming session. By contrast, for a diver more than 50 metres (164 feet) out in cool waters more than 5 metres (16 feet) deep in springtime, when mother whales and their calves are present, the risk goes up considerably to one in 15,000, which means that a bather or surfer can make informed choices about the risk before taking to the water. They have to be particularly alert to the dangers

during winter and spring. It also means, of course, that Western Australia's intended *Jaws* Effect cull of large white sharks
using drumlines in the summer of 2013–14 was a complete
waste of time.

On the other side of the world, Cape Cod on the US east
coast has had a sudden influx of white sharks because the population of seals has increased exponentially. Unlike in many
white shark hotspots, where the seal rookeries are on offshore
islands, such as Geyser Rock in South Africa and Neptune
Island off South Australia, Cape Cod's seals share the dunes
with beachgoers. There have been several shark bite incidents
and people have lost their lives, but the local community has,
to some extent, embraced the presence of the sharks rather
than fight them. Local businesses, for example, display and
sell T-shirts, towels and hats with bold shark designs. In 2015,
according to a report in the *Boston Magazine*, the Chatham
Merchants Association opened its outdoor art exhibition
Shark in the Park, and the Chatham Orpheum Theatre has,
since 2013, shown *Jaws* during the week of the 4 July holiday,
with cinemagoers shouting out the words of the dialogue and
crushing beer cans like Quint.

However, not all residents are so keen to embrace their new
neighbours. The *Jaws* Effect has infiltrated here too. All are not
agreed on what do about the sharks. One regional commissioner advocated killing all the sharks. He proposed a shark
hazard mitigation strategy – drumlines off popular beaches
and, if the hooked shark is not dead, then it should be shot
– sound familiar? However, having backtracked on killing
sharks because of the pressure from activists, the commissioner advocated a cull of the sharks' main food – seals. And, with

their main food supply gone, what are the sharks going to feed on? Right, you got it.

While these kind of revenge attacks on white sharks can only be sanctioned by the authorities nowadays because they are protected in many of the world's white shark hotspots, such as Australia, South Africa, California, Massachusetts and Nova Scotia, it wasn't always that way. Directly after the *Jaws* book and movie were released, the macho thing to do was to catch a white shark, string it up at the quayside to show how brave you were, and then cut out the jaws, either displaying them, like the rows of jaws in Quint's fishing shack, or selling them to the highest bidder. Either way, hundreds, if not thousands, of white sharks perished. 'The film *Jaws* triggered a collective testosterone rush amongst fishers,' said George Burgess, who was once director of the Florida Program for Shark Research and curator of the International Shark Attack File. 'Thousands of fishers set out to catch trophy sharks after seeing *Jaws*,' he told the BBC. 'It was good blue collar fishing. You didn't need to have a fancy boat or gear – an average Joe could catch big fish, and there was no remorse, since there was this mindset that they were man-killers.'

He went on to point out that 'on North America's east coast, the number of large white sharks declined immediately by close to 50 per cent'. This was borne out by research on shark populations in the north-west Atlantic Ocean by Julia Baum and colleagues from Dalhousie University, Halifax, Nova Scotia, who revealed that, a little later between 1986 and 2000, there were estimated population declines of 75 per cent of white sharks, tiger sharks and scalloped hammerhead sharks. Much of this was due to overexploitation by commercial fisheries, but until 1997 in US federal waters and 2005 in Massachusetts state waters, when white sharks gained full protection, sports fishermen accounted for a good few individuals.

It was not helped by the bizarre notion that sharks would make for a good competition – the so-called shark tournaments, killing sharks for fun not food. The biggest sharks won, and had to be weighed ashore, so inevitably large white sharks ended their days strung up at the quayside in front of gawping crowds. After *Jaws*, shark tournaments suddenly sprang up in California, and what with a renewed commercial interest in catching sharks, which resulted in overfishing, white sharks off the US west coast were almost exterminated.

By 2016, on the east coast, there were seventy-two shark fishing tournaments registered with NOAA Fisheries for pelagic fisheries in the US Atlantic Ocean, Gulf of Mexico and Caribbean Sea, and they were responsible for the killing of up to 70,000 sharks, including white sharks, in a single fishing season. However, with a more enlightened public, only fourteen tournaments exist today, and they cannot land white sharks. Many events lost their corporate sponsors. New England, for example, closed down its last tournament a few years ago; so the tide *is* turning. There *is* hope for this magnificent creature.

When mention is made of the great white shark, it's hard not to have the ominous two-note theme to *Jaws* playing in the back of one's mind – but more frequently nowadays, it's being replaced by a fresh, more positive and less scary tune.

White sharks are beginning to throw off their old man-eater image and, as we have seen in the chapters of this book, researchers are revealing them to be complex, intelligent and essential members of ocean communities. Sharks, including white sharks, are crucial to the functioning of marine ecosystems. They mop up the dead and dying, thus confining the spread of diseases, and they ensure that prey animals do not overpopulate and upset the food chain. All in all, sharks, and white sharks in particular, are remarkable animals. Peter Benchley became a fan. 'Sharks', he said, 'are nature at its most perfect, the most beautiful thing in the water you can imagine.'

And now there are international agreements limiting their exploitation, as well as improvements in the way catches are recorded and the safe release of sharks that have been caught accidentally. It's not perfect, but it's a start.

Scientists are also focusing on the interface between white sharks and humans, making water sports safer for swimmers and surfers and, while keeping them safe, reducing the number of sharks that are killed unnecessarily. Drones, which have had such negative publicity in war zones like Ukraine, are having a positive impact on safety at beaches. Modern technology is finally catching up with an animal that has its roots in events that occurred several million years ago.

There are also ways in which members of the public can make a contribution. In the UK, for example, you can join the Shark Trust, which 'highlights the problems faced by sharks and taking those problems to find real-world solutions', and you can get involved by taking part in one of the trust's citizen science projects. It won't see you coming face to face with a white shark, but it will contribute to the wellbeing of sharks generally. The website www.sharktrust.org is easy to access. However, for the last word, maybe we go back to the beginning: *Jaws*, which is celebrating its 50th anniversary in 2025. Its director reflected on the film's impact.

Speaking in October 2024, on BBC Radio 4's *Desert Island Discs*, Stephen Spielberg was asked by the show's presenter, Lauren Laverne, how he would feel if he was on a desert island surrounded by shark-infested waters. He replied:

> That's one of the things I still fear – not to get eaten by a shark, but that sharks are somehow mad at me for the feeding frenzy of crazy sports fishermen that happened after 1975. I truly, and to this day, regret the decimation of the shark population because of the book and the film.